QH 588 .S83 S35 2006 00005014 7
Stem cell now :
W9-BZV-867

A PLU

STEM C

CHRISTOPHER THOMAS SCOTT is executive director of the Stem Cells in Society program at the Stanford University Center for Biomedical Ethics. He was formerly the assistant vice chancellor at the University of California, San Francisco, and was a founder and the executive editor of the award-winning biotech journal *Acumen*. He has appeared on national radio and television, and has written for major newspapers and journals such as *Science, Nature Biotechnology,* and *The Scientist*.

"The image on the cover of this book reminds me of Carl Sagan's Pale Blue Dot. Like Planet Earth, a single stem cell may well prove to be the starting point of a long and cosmic journey for humankind. Anyone who is interested in this journey (and its fascination should be universal) will find *Stem Cell Now* a solid and useful guide."

—Jonathan Weiner, Pulitzer Prize–winning author of *The Beak of the Finch*

"By reducing complexity to simplicity without losing meaning, Scott has provided a firmly grounded and well-illustrated schooling in our current understandings and misunderstandings of how stem cell research may play out in creating new forms of therapy for currently untreatable diseases. Scott enables readers to assess more meaningfully and effectively the conflicting assessments of the medical promise and ethical concerns that have dominated the public debates."

—Paul Berg, Nobel laureate, Stanford University, and coauthor of *Dealing with Genes*

"In this engaging book, biologists give their first-hand accounts of why their work generates such passion and attention. If you need an introduction to the intricacies of stem cell science and some of the complex ethical arguments behind the debates, this is a fine place to begin."

—Laurie Zoloth, Director of the Center for Bioethics at Northwestern University and coeditor of *The Human Embryonic Stem Cell Debate*

WESTWOOD COLLEGE-LIBRARY
RIVER OAKS CAMPUS

STEM CELL NOW

A Brief Introduction to the Coming Medical Revolution

CHRISTOPHER THOMAS SCOTT

A PLUME BOOK

PLUME
Published by Penguin Group
Penguin Group (USA) Inc., 375 Hudson Street, New York, New York 10014, USA • Penguin Group (Canada), 90 Eglinton Avenue East, Suite 700, Toronto, Ontario, Canada M4P 2Y3 (a division of Pearson Penguin Canada Inc.) • Penguin Books Ltd., 80 Strand, London WC2R 0RL, England • Penguin Ireland, 25 St. Stephen's Green, Dublin 2, Ireland (a division of Penguin Books Ltd.) • Penguin Group (Australia), 250 Camberwell Road, Camberwell, Victoria 3124, Australia (a division of Pearson Australia Group Pty. Ltd.) • Penguin Books India Pvt. Ltd., 11 Community Centre, Panchsheel Park, New Delhi – 110 017, India • Penguin Group (NZ), 67 Apollo Drive, Rosedale, North Shore 0632, New Zealand (a division of Pearson New Zealand Ltd.) • Penguin Books (South Africa) (Pty.) Ltd., 24 Sturdee Avenue, Rosebank, Johannesburg 2196, South Africa

Penguin Books Ltd., Registered Offices: 80 Strand, London WC2R 0RL, England

Published by Plume, a member of Penguin Group (USA) Inc. Previously published in a Pi Press edition.

First Plume Printing, September 2006
10 9 8 7 6

Copyright © Christopher Thomas Scott, 2006
All rights reserved

 REGISTERED TRADEMARK—MARCA REGISTRADA

The Library of Congress has catalogued the Pi Press edition as follows:

Scott, Christopher Thomas.
Stem cell now : from the experiment that shook the world to the new politics of life / Christopher Thomas Scott ; foreword by Donald Kennedy.
p. cm.
Includes bibliographical references and index.
ISBN 0-13-173798-8 (hc.)
ISBN 978-0-452-28785-3 (pbk.)
1. Stem cells. 2. Embryonic stem cells. 3. Human reproductive technology. I. Title.
QH588.S83.S35 2006
616'.027774—dc22 2005023266

Printed in the United States of America

Without limiting the rights under copyright reserved above, no part of this publication may be reproduced, stored in or introduced into a retrieval system, or transmitted, in any form, or by any means (electronic, mechanical, photocopying, recording, or otherwise), without the prior written permission of both the copyright owner and the above publisher of this book.

PUBLISHER'S NOTE
The scanning, uploading, and distribution of this book via the Internet or via any other means without the permission of the publisher is illegal and punishable by law. Please purchase only authorized electronic editions, and do not participate in or encourage electronic piracy of copyrighted materials. Your support of the author's rights is appreciated.

BOOKS ARE AVAILABLE AT QUANTITY DISCOUNTS WHEN USED TO PROMOTE PRODUCTS OR SERVICES. FOR INFORMATION PLEASE WRITE TO PREMIUM MARKETING DIVISION, PENGUIN GROUP (USA) INC., 375 HUDSON STREET, NEW YORK, NEW YORK 10014.

For Mom and Dad

Contents

Foreword

One of the pleasures I have had as editor-in-chief of *Science*—the weekly journal of the American Association for the Advancement of Science—has been to watch the extraordinary progress of biomedical science. No branch of science has grown more rapidly or produced more dramatic results than biomedicine. Examples include the sequencing of the human genome, "genetic engineering" using recombinant DNA, and the molecular biology of memory and mood disorders. Biomedicine in the United States has become largely an enterprise of government: the majority of the funds are supplied by the National Institutes of Health (NIH), whose multibillion dollar annual budget supports an army of university researchers and the technological and chemistry industries that supply it.

It cannot be a surprise, then, that progress in biomedicine can swiftly become political. The government that supplies the resources also makes rules about what can be done with them. Thus, for example, NIH requires that its committees approve new experiments that will make use of recombinant DNA techniques, and mandates that universities receiving its funds add to their curriculum courses in the ethical conduct of research for graduate students. In *Stem Cell Now,* Christopher Scott tells the story of how an extraordinary scientific breakthrough—one with potential future uses in the treatment of disease—developed a political overlay that has become a major issue for the Congress and the president of the United States.

A central element in the political drama has been the public's confusion about different terms that play pivotal roles. One such term is "cloning." As Scott explains, cloning is popularly associated with an experimental process by which an animal is induced to produce a genetically identical copy of itself. That's how Dolly the sheep gained celebrity status—she was produced by injecting the nucleus of a skin cell from one sheep into an enucleated egg cell from another. The technique that enabled this experiment to succeed (most failed) is called somatic cell nuclear transfer.

In entirely different experiments, Scott shows how embryonic stem cells can be isolated from a mammalian embryo after it has undergone a few divisions. These cells can divide repeatedly in culture, producing what is called a "line" of stem cells. They are clones of one another, but when injected into an adult organism they are able to differentiate to form—or replace—damaged cells of the adult tissue in which they reside. The latter process is sometimes called therapeutic cloning, because the intended use of the technology is to repair defective tissue. A different process aimed at reproduction made Dolly, and hence is sometimes called reproductive cloning.

The public confusion between these two has agitated the politicians and rattled the scientific community. There is widespread fear that reproductive cloning of humans is in our future, and a few publicity-seekers have claimed, without supplying evidence, to have done so. No scientist I know would support such a use, and it enjoys little support from philosophers. But in bills pending before the Congress, the fundamental technique of somatic cell nuclear transfer—potentially useful in basic research on very early human development—is made subject to criminal penalties including a ten-year jail term. In the meantime, the ban on using federal funds to create new stem cell lines, established by President Bush's order in 2001, still holds.

Dispelling the confusion is important to the future of this work, and Christopher Scott—a scientist with a journalist's skill at clear explanation—has provided exactly what is needed in this book. It is clear that Scott has a position on the potential value of stem cell research, and on the political division that swirls around it. Yet he gives a reliable, balanced, and thoughtful account of the biology of stem cells and the history of this remarkable new advance in our understanding of the process of development.

Recent events have only underscored the importance of this issue, as well as of the high stake politics that surround it. It is plain that powerful support has developed in favor of ending the ban on creating new stem cell lines, and in the early summer of 2005 both the Senate and the House of Representatives passed bills to that effect—despite the promise of a presidential veto. Just days earlier, *Science* had published a widely hailed paper from a group of South Korean scientists. It reported that eleven new human embryonic stem cell lines had been made using eggs from women who had given informed consent as volunteer donors. The Korean team, already having perfected its somatic cell nuclear transfer technique first in animals and then in humans, used nuclei from somatic cells of individuals with different congenital diseases. The team appeared, at the time, to have achieved the first large-scale success at producing multiple new lines.

To everyone's discomfiture, obviously including mine, that work has now been demonstrated to have been fraudulent—indeed, one of the most comprehensive and convincing frauds in the history of research misconduct. Scott recounts in Chapter 10 how the debacle aroused intensive press coverage, raised questions about the reliability of the peer review process, and produced a thorough study at *Science* designed to explore that question. In the aftermath, many have wondered whether the South Korean scandal will affect the prospects for further work on stem cell biology in the United States. So far, the reactions have been relatively muted, apart from a few statements decrying the fraud itself. Leading investigators in the U.S. and elsewhere have continued their work and indicate optimism about the future of stem cell research.

As Scott explains in Chapter 4 the nuclei from the somatic cells, although they are "reprogrammed" by the egg cytoplasm to permit normal early development, eventually exert their own genetic profile on the resulting stem cells. One objective will be to succeed where the Korean group failed: to develop stem cells that have the immunological signature of the donor nucleus, so that they could potentially be used to treat the donor who has the disease without undergoing rejection. Of course, such therapeutic applications remain in the distant future, and Scott is appropriately cautious about premature promises that raise unjustified hopes. Nevertheless, major advances in the field get instant attention.

Resolution of the ethical and religious objections to stem cell research will doubtless take some time. If our society is to reach a consensus of sorts on this issue, a substantial growth in public understanding will be required. For that, a careful and clear explanation of the science, its history, and the policy challenges it presents is essential. That account has not been available until now, to the disappointment of the scientific community and the frustration of thoughtful citizens who are looking for answers. This thoroughly responsible and literate effort should fill that need.

Donald Kennedy
Editor-in-Chief, Science *Magazine*

Abbreviations

ESC	embryonic stem cell
FDA	Food and Drug Administration
GVHD	graft-versus-host disease
hESC	human embryonic stem cell
HHS	Health and Human Services
HLA	histocompatibility antigens
HSC	hematopoietic (blood-forming) stem cell
ICM	inner cell mass
IVF	*in vitro* fertilization
MAPC	multipotent adult progenitor cell
NAS	National Academy of Sciences
NIH	National Institutes of Health
NSC	neural stem cell
PCBE	President's Council on Bioethics
PGD	pre-implantation genetic diagnosis
SCID	severe combined immunodeficiency
SCNT	somatic cell nuclear transfer (or nuclear transfer)

STEM CELL NOW

1

The Experiment That Shook the World

The development of cell lines that may produce almost every tissue of the human body is an unprecedented scientific breakthrough. It is not too unrealistic to say that this research has the potential to revolutionize the practice of medicine and improve the quality and length of life.[1]

FORMER NIH DIRECTOR AND NOBEL PRIZE WINNER
HAROLD VARMUS

In the November 6, 1998, issue of the journal *Science*, James Thomson, a professor at the Wisconsin Regional Primate Research Center at the University of Wisconsin, reported he had developed the first line of human embryonic stem cells. Penned in the typical understatement of research writing, the abstract of the research report declares, "These cell lines should be useful in human developmental biology, drug discovery, and transplantation medicine."[2] Depending on one's philosophical bent, the implications of this statement were momentous—or disastrous. An incredibly potent human cell was alive and well, living in an incubator in James Thomson's laboratory.

Only three pages long, the Thomson paper is packed with information and data. He describes how he obtained human embryos from a local *in vitro* fertilization (IVF) clinic. Couples were given the option of donating extra embryos for research purposes—they were left over from the IVF procedure. They arrived packed in ice, frozen just days after fertilization in a laboratory dish. Visible only under a microscope, each embryo contained about eight cells surrounded by a very thin membrane—resembling a diaphanous sac with a cluster of soap bubbles inside. Thomson placed the transparent orbs into culture dishes with carefully prepared nutrients and grew them into blastocysts, hollow spheres of about 100 cells, as shown in the figure. At this stage the embryo is between four and five days old and scarcely a tenth of a millimeter across, about twice the diameter of a human hair. Inside the cavity of the blastocyst is a mound of cells called the inner cell mass, or ICM. With a microscope, a steady hand, and a very thin, hollow glass needle, Thomson removed the clump of cells from inside the sphere and placed them in a laboratory culture dish.

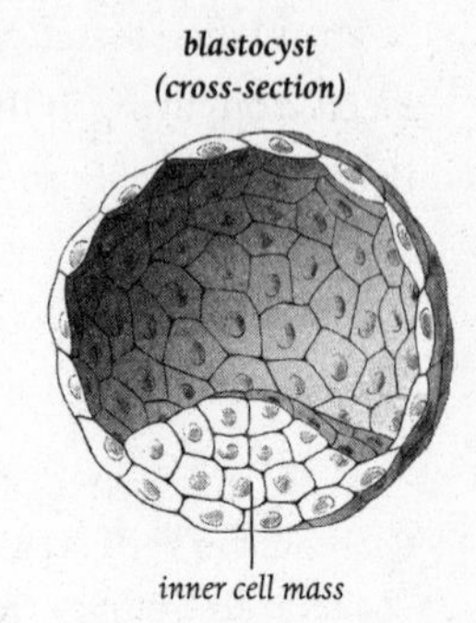

Now came the difficult part: how to ensure the cells would live and thrive in the laboratory. Cell culture, the process by which biologists grow living cells in a plastic culture dish, is a tricky and time-consuming business. Cells must grow in a sterile environment, or airborne contamination will ruin the experiments. Getting the growth nutrients just right is another hurdle. The conditions in a cell culture dish (or *in vitro* from the Latin *in glass*) must essentially replicate the environment of a cell growing and dividing inside the body (or *in vivo* from the Latin *in life*). Other challenges include maintaining the right temperature, the right oxygen and carbon dioxide concentrations, and deciding when to change and refresh the growth nutrients. With some trial and error, Thomson's cells began to multiply. Fortunately, they also had the staying power to persist in this artificial environment for extended periods of time. In fact, when his paper was published, Thomson's cells were still robust and had survived and multiplied for eight months. The longevity of his cell lines was the first of three groundbreaking results reported in his *Science* paper.

The second result was just as crucial. The cells demonstrated no ill effects from living in their laboratory environment. Cells proliferate by dividing in two. Sometimes artificial conditions cause cells to divide incompletely, or not at all, or such conditions result in abnormal numbers of chromosomes, which are the structures that contain the cell's genes. The ingredients in the culture dish can also have ill effects on the genes themselves, causing the cell to change physical characteristics or to age prematurely and die. It was important that Thomson's cells remain consistent in type and function, even after dividing many times over many months, so they could be used reliably in future experiments.

If an apparently consistent culture of cells dividing rapidly for many months was this experiment's "lightning," the third result was its thunderclap. With careful manipulation, laboratory technicians removed the human cells from the culture dishes and injected them into experimental mice. These mice are engineered to lack an immune system so they do not reject the human cells. Once in the mice, the human cells divided rapidly and formed tumor-like structures made up of all the major human tissue types, including skin, muscle, and bone. The cells bore the unmistakable imprimatur of embryonic stem cells—next to the fertilized egg itself, the most powerful cells in the body. The thunderclap then was what Thomson showed: stem cells can be coaxed into becoming any tissue type in the human body.

The scientific and medical implications contained in this short paper are profound and unambiguous. Embryonic stem cells could be used to generate new tissue and organs for transplantations. Defective and dying tissues caused by diseases such as Parkinson's or diabetes could be replaced with an unlimited supply of specially grown stem cells. Cultures of human stem cells could be used as laboratory tools to help identify new drugs and therapies. For pure scientists like Thomson, observing stem cells in the laboratory could provide insights into how all animals embark upon the magnificent developmental process that begins with a single cell.

A RECIPE FOR SUCCESS

When the 1998 *Science* article rolled off the press, James Thomson was just 38. Originally trained as a biophysicist, he had a doctorate in veterinary medicine and a Ph.D. in molecular biology. For any research biologist, a *Science* publication is a significant achievement. But it was far from Thomson's first paper. He already had twenty others listed on his resume.

Thomson started his graduate career at Pennsylvania's Wistar Institute in the mid-1980s under the wing of an early pioneer of developmental biology, Davor Solter. Known for his work on mouse embryology, Solter took a blastocyst from the animal's uterus, teased it apart, and placed the cells of the ICM in special culture conditions that allowed them to survive and multiply. What Thomson learned from his training with Solter was pivotal to his later years of research. The recipe for a specially designed laboratory "soup" (or in the parlance of biologists, the *medium*) into which embryonic cells are placed to grow and divide means the difference between cells that thrive and cells that wither. Through trial and error, Solter perfected the ingredients in his mouse embryonic cell medium—it worked so well that the basic recipe is still widely used today.

During the interim period between his graduate thesis defense and the next stage of his training (called a postdoctoral fellowship), Thomson continued his work with mouse embryonic stem cells. At the Roche Institute of Molecular Biology, he teamed up with stem cell expert Collin Stewart. At that time, embryologists were trying—unsuccessfully—to grow human embryonic cells using what they had learned from mouse embryonic cells. During lunch with Stewart in 1988, Thomson had an epiphany. "Collin told me about people in Britain who had attempted to derive human embryonic stem cells but had failed," he recalls.[3] "The problem became obvious. If you compare a mouse embryo to a human embryo, they are as different as night and day. Even some of the molecules that control the embryo's development in the mouse are different or missing entirely in humans."[4] He reasoned that if he could work out the cell culture recipe on a species closely related to humans, he would be one step closer to solving the scientific hurdles blocking human embryonic stem cell research.

At the tender age of 30, Thomson did just that. He went off to do his fellowship at the Oregon Regional Primate Center, which was at that time the best training ground for primate biologists. While there, he perfected his cell culture techniques and, in 1991, he was recruited to the University of Wisconsin to work on monkey embryonic stem cells. Four years later, he derived the first primate embryonic stem cell line and, in 1995, published his research in the *Proceedings of the National Academy of Sciences of the USA*.[5] As the paper went to press, Thomson held his breath. "After the monkey research was published, I fully expected other labs to use our methods to do the same thing with human cells," he said. "But no one did." Two years later, Thomson figured he would try it himself. "It was surprisingly easy," he recalls. "We had worked out most of the techniques already." He paused for a moment. "You know, there is a certain amount of finesse to growing the cells of this type, and most of our failures came with the monkey stem cells. It was worth the time: the very first human stem cell we isolated gave us a cell line!"

Although the human health implications of a line of human stem cells were not lost on him, Thomson focused on the mechanics of an animal's development: how genes orchestrate the process, what chemical signals are involved, and how the combination leads to organized structures such as skin and bone. He knew that a system to grow embryonic stem cells would be used as a standard tool for other biologists, and as a result, the entire field would benefit. Ted Golos, a fellow faculty member at the University of Wisconsin and collaborator on the monkey research, describes Thomson as a "how things work" kind of scientist. Golos says, "It can be dangerous if your interests don't have immediate benefit to solving a human disease because the government sometimes doesn't fund 'how things work' kind of projects."[6]

Advances from the reproductive biology field aided Thomson's success with human cells. After fertilizing a human egg in a test tube, an IVF clinician incubated it for a brief amount of time before placing it back into the mother. Early procedures met with limited success. The cell culture medium was unable to mature the fertilized egg to an age where it could "take on" the environment of the uterus and survive. As a result, doctors transferred embryos too early, resulting in what Thomson calls "developmental mismatches." By the mid 1990s, the culture medium had improved markedly, along with the rate of successful IVF pregnancies. Coincidentally, when Thomson switched to human embryonic stem cell research, the new media became available. Then Thomson recruited a postdoctoral fellow who had trained with the inventor of the new medium and adapted it to his own methods.

THINGS HEAT UP

Thomson and his colleagues, like most research biologists, are part of an international network of scientists working in universities, research institutes, and corporations. Since 1945 American universities with biological and medical sciences programs have benefited from the bounty of the Department of Health and Human Services (HHS) and its biggest agency, the National Institutes of Health (NIH). It is a foregone conclusion that without the NIH, the National Science Foundation, the Department of Energy, and other government agency funding, Americans would not enjoy some of the best medical care in the world. Although the brute force of government spending on biological science hasn't always yielded immediate results, by most measures, America has benefited greatly from the investment. Americans have access to a powerful repository of drugs, therapies, and medical devices—a dizzying array of technology designed to propel us into a happy and healthy old age.

But not all has been congenial between biomedical scientists and their funding agencies. Presidents and Congresses, both liberals and conservatives, have used their authority to guide, redirect, and limit funds. In this aspect, the fate of science funding is no different than funding for interstate highway systems, municipal police departments, or the National Endowment for the Arts. As it turned out, James Thomson and other human embryologists did their work not with government resources but with private funds. Why? Because government support of research using human embryos has been banned by Congress for decades. The controversy began in the late 1970s with the advent of IVF and the spare embryos generated by the procedure. Most proponents of biomedical research hold that it is morally permissible, even morally required, to use the extra embryos for potentially life-saving biomedical research. Opponents object, saying that the destruction of any embryo is the moral equivalent of killing a human life.

Soon after the Thomson paper was published, the NIH, recognizing the potential of human embryonic stem cell research, sought to lift the congressional ban, and NIH director Harold Varmus said he would draft guidelines regulating the use of embryonic cells. In 1999, President Clinton asked for a review of the matter by his ethics experts, and they concluded that the federal government should fund research provided that only embryos left over from fertility treatments be used. The recommendations clearly stated that the parents must have donated the embryos expressly for the research and that the IVF clinics must not profit from the exchange.

That year, *Science* proclaimed the development of human stem cell lines as the most important advance of the year. Cn its annual top ten list, it said, "In just one short year, stem cells have shown promise for treating a dizzying variety of human diseases." Similar reports followed from the major media outlets. CNN breathlessly reported, "Researchers isolate human stem cells in the lab; breakthrough could lead to treatments for paralysis, diabetes."[7] Amidst the commotion, however, were growing criticisms and warnings from religious and moral leaders. The National Council of Catholic Bishops protested, calling the White House policy to allow the use of otherwise-discarded early embryos "guidelines on how to ethically destroy human life."[8] Pope John Paul II weighed in, calling stem cell research an "accommodation and acquiescence in the face of other related evils, such as euthanasia [and] infanticide." He went on to say, "A free and virtuous society, which America aspires to be, must reject practices that devalue and violate human life at any stage from conception until natural death."[9] The fact that cloned animals were now part of the scientific scene muddied waters further—the procedure used to make Dolly the sheep shares its scientific history with embryo research. Scientists and journalists used words such as embryo and cloning so cavalierly that the lay public wasn't sure what distinguished animal cloning from babies conceived through IVF and embryonic stem cell research. As the millennium drew to a close, many people felt that a knock on the door from their human clones seemed a distinct possibility.

Clinton signed the NIH guidelines in August 2000, opening the door to scientists who needed funding for embryonic stem cell research. But few were willing to risk the precious time to write grant proposals that could be rescinded with sudden shifts in political winds. During a campaign speech, George W. Bush made clear his intentions regarding the issue, saying, "I oppose federal funding for stem cell research that involves destroying living human embryos."[10] In an election year riddled with controversy, the stage was set for a raging battle in which scientists, politicians, religious leaders, doctors, and patients would find themselves unwilling soldiers.

One year later from his ranch in Crawford, Texas, President Bush made a sweeping announcement: funding from the NIH would be used for research only with preexisting embryonic cell lines (which numbered only in the dozens), and no federal funds would be used for the creation or use of new stem cell lines or to clone human embryos for any purpose. Later that same year, the House of Representatives followed the administration's lead and, by a wide majority, banned cloning of humans and voted to criminalize so-called therapeutic cloning, a laboratory method used to generate embryonic stem cells. The penalty was set at a $1 million fine and up to ten years in jail. In January 2002, the Senate swung into action, and Sam Brownback (R, Kansas) introduced a proposal that mirrored the House's bill. The Senate failed to act on the legislation in 2003 and 2004. In 2005, momentum in favor of stem cell research began to swing slowly the other direction. In a challenge to President Bush, the House of Representatives approved legislation to lift the ban on embryonic stem cell research. The vote was 238 to 194, 47 votes short of the two-thirds majority needed to override a presidential veto. "I made it very clear to the Congress that the use of federal money, taxpayers' money, to promote science which destroys life in order to save life—I'm against that," Bush said before the vote. "And therefore, if that bill does that, I will veto it."[11]

ADULT VERSUS EMBRYONIC STEM CELLS

As the debate escalated, opponents of embryonic stem cell research mined emerging scientific evidence suggesting that adult stem cells could be used in therapy. Although they show up prior to birth, adult stem cells are developmentally older, specialized cells that exist in many places in the body, biding their time before they replace old and damaged cells and the diseased tissues in which they reside. The principal difference between adult and embryonic stem cells lies in their potential to become different types of cells and tissue. Embryonic stem cells have enormous potential—they can become any cell or organ in the body. Adult stem cells, by contrast, are restricted in what they can become. As they mature, their ability to change becomes increasingly limited until they are a fully matured cell, such as a skin cell or a neuron.

Preliminary results from some adult stem cell research laboratories in late 1999 and early 2000 hinted that adult stem cells were every bit as powerful as their embryonic counterparts. Political and religious groups used this data to make the case that embryonic stem cell research was unnecessary. But other laboratories were unable to repeat the experiments, and a burst of new data refuted the original claims. As a result, few stem cell researchers today will say that adult stem cells are the sole answer for curing disease and physical dysfunction. Indeed, experts say the opposite—embryonic stem cells have tremendous therapeutic potential.

The back and forth about the power of adult stem cells pales in comparison to what became the biggest scientific scandal in years. In 2005, a South Korean scientist, Hwang Woo Suk, claimed to have made therapeutically useful lines of embryonic stem cells using the cloning technique that produced Dolly the sheep. The success was the first convincing evidence that the method could someday be used to treat disease. Months later, whistle-blowers and investigative journalists began to question his results, and by January 2006, a probe by Hwang's university proved he had fabricated his experiments. The fraud dealt stem cell research a significant blow, and some wondered whether embryonic stem cells would ever live up to their promise.

The final answers won't come any time soon. Cures may come from other disciplines of biology and some diseases may prove too stubborn to treat. Human stem cell research is in its infancy and is extremely fluid: results published in last year's scientific journals are quickly refuted this year. Biology yields its secrets grudgingly, and it is quite possible that a decade or more will pass before anything resembling a general theory of stem cell biology is articulated, and then only if research is allowed to proceed under open conditions. Our knowledge of human development relies on investigating both types of cells: prohibiting one line of research and not another is like asking Einstein to understand relativity without gazing at the stars or asking da Vinci to understand flight without watching birds.

A GLIMPSE OF WHAT LIES AHEAD

What are stem cells? Can they cure diabetes or make a new heart? What is the scientific controversy that swirls around embryonic and adult stem cells? Which stem cell discoveries will develop into actual therapies? Which diseases will benefit, and how long will it take? If we aggressively pursue embryonic stem cell research, will human clones be far behind? Do we destroy a human being when we use an embryo for research?

The pages ahead provide the biological and scientific basics needed to explore the most recent advances and therapeutic applications of human stem cells. But a glimpse into biology and medicine provides only part of the answers. No area of science is so deeply interwoven with ethical concern. The final chapters cover the moral and political dimensions that, along with their medical promise, make stem cells front-page news. It is difficult to find a biologist who will say that stem cells alone hold the key to solving our most intractable diseases. But it is safe to say that no single area of biomedicine holds such great promise for improving human health.

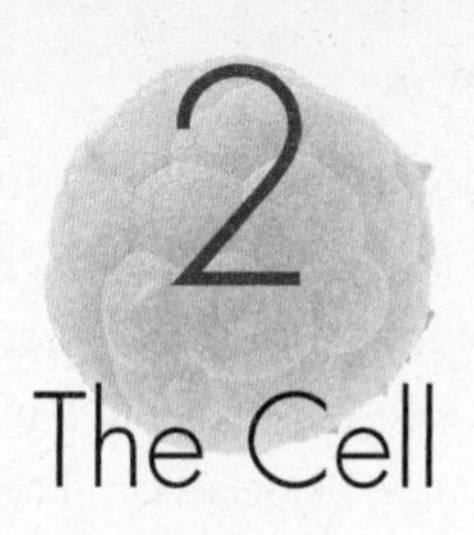

2 The Cell

I could be bounded in a nutshell and count myself a king of infinite space.
WILLIAM SHAKESPEARE, HAMLET

If a book has the word *cell* in its title, it had best start with the small specks seen only with a microscope. Looking deep into this minute and silent world, we see that cells are complex living units. They contain the basic instructions for living and thriving, the genes. Genes, the discrete units of inheritance, determine every characteristic of a living thing: how a creature looks and behaves; how it grows and thrives. The most powerful feature of nearly all cells is their ability to create identical copies of themselves in order to pass their genetic information on to the next generation. The ability to self-replicate figures prominently in the following chapters, because cell division is the central mechanism behind stem cells and the developing animal. Once the basics of cell biology are in hand, the particular magic of stem cells becomes abundantly clear.

MAGNIFICENT MICROCOSMS

Cells are the basic structural units of all organisms, from a one-celled bacterium to a multitrillion-celled human. Most animal cells have common features and operating styles. The basic cell characteristics can be seen in this illustration of a white blood cell, or immune cell. Observed with a microscope, a cell looks like a water-filled balloon or sac. A thin layer of fat called a plasma membrane surrounds the cell like diaphanous skin. The plasma membrane regulates what enters and what leaves the cell. Some things, like small molecules, can move through the membrane, and others, such as large molecules and important structures in the cell's insides, cannot. It is the "insides," or jelly-like cytoplasm, where most of the action takes place. In typical animal cells, the largest structure inside a cell is the nucleus, another sac with its own membrane. (An exception is red blood cells, which lack nuclei.)

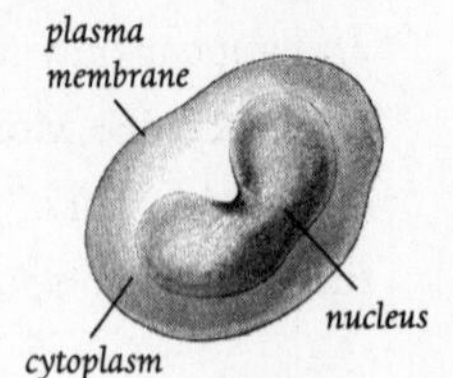

white blood cell

Humans are made of hundreds of different kinds of cells, a few of which are shown here. The red blood and immune cells circulate throughout the body. By contrast, bone cells are mostly static. Nerve cells, or neurons, are part of a diffuse matrix and can stretch over great distances (microscopically speaking). Other cells are localized and tightly packed in organs and

red blood cells (lacking nuclei)

bone cell

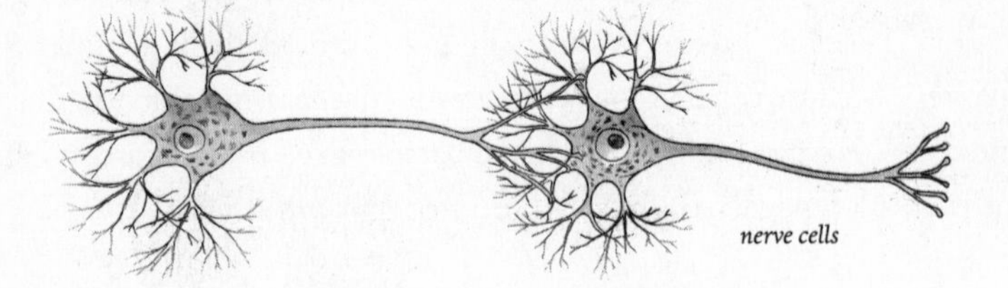
nerve cells

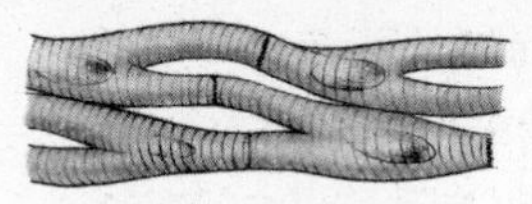
heart muscle cells

tissues, pulling together as a functional whole. For example, muscle cells of the heart exhibit a steady synchrony of response that is astounding in its staying power. Some cells stay with the organism from birth until death; others are ephemeral and appear only for a particular function, designed to be replaced by new recruits. Most cells don't do their work in isolation—they respond to the animal's demands by communicating with each other by chemical signals delivered through their plasma membranes.

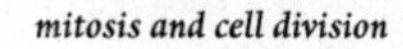
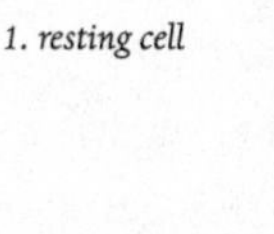
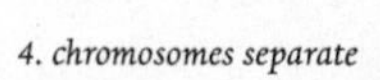
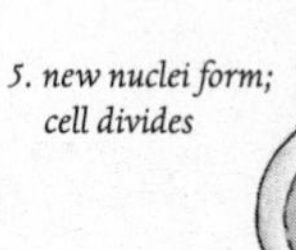
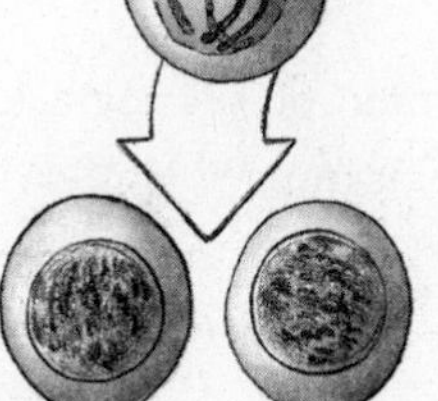
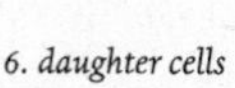

mitosis and cell division

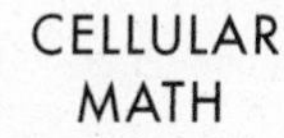

CELLULAR MATH

Cells multiply by dividing. That is, they make copies of themselves with a burst of activity that splits them in two. The action begins in the nucleus, which contains chromosomes, threadlike structures that are the repositories for genes. In humans, each body cell contains 46 chromosomes—found in 23 pairs—one set inherited from the mother, the other from the father. During cell division, these chromosomes duplicate and split into two new cells, each containing the same number of chromosomes as the initial cell. This chromosomal dance, illustrated on the previous page, is highly orchestrated. First, the long, filamentous chromosomes begin to double and condense into tightly wound bundles. The nuclear membrane surrounding the chromosomes becomes less distinct and disappears altogether.

Then, tubular scaffolding attaches to each of the chromosome pairs. The forty-six chromosomes, now distinctly twinned, are pulled into a line. (For simplicity, only four chromosomes are shown in the drawings.)

Next, the chromosomes separate and move to opposite sides of the cell. The new chromosomes elongate, relax, and unwind—and perhaps enjoy a toast. Two new nuclear membranes form, each containing an identical set of chromosomes. The cell begins to pinch in two and, minutes later, two cells exist where one used to be. Mitosis—the term for replication and division of chromosomes in the nucleus—and cell division have produced a daughter cell, a genetic clone, identical to its parent. In humans, cell division takes about twelve hours.

This ability to divide serves several purposes for stem cells. For an animal to grow and form larger structures, such as tissues and organs, stem cells need to produce large quantities of new cells, sometimes on rather short notice. In the mature organism, stem cells replace and replenish cells that are injured as well as those that have grown old and died.

DNA, GENES, AND GENOMES

The chromosomes in a nucleus are tangles of DNA, wound like yarn kept by a deranged knitter. In turn, DNA is fashioned into a twisted ladder (called a helix) built of two strands, each containing a quartet of molecules called nitrogenous bases, linked one after another and

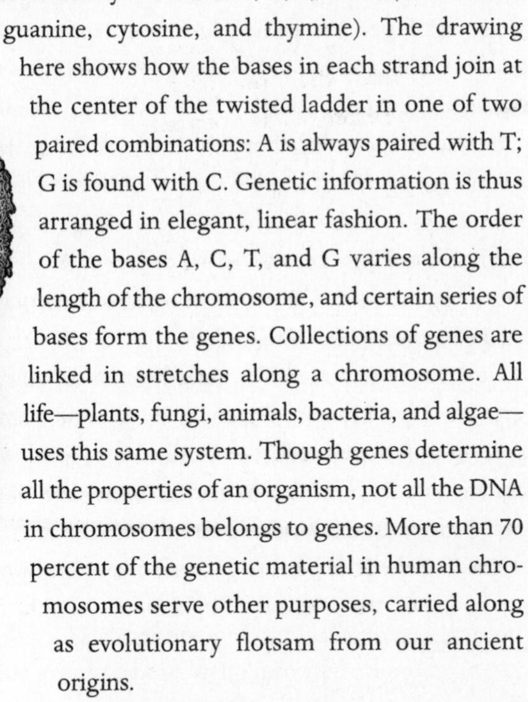

abbreviated by the letters A, C, T, and G (for adenine, guanine, cytosine, and thymine). The drawing here shows how the bases in each strand join at the center of the twisted ladder in one of two paired combinations: A is always paired with T; G is found with C. Genetic information is thus arranged in elegant, linear fashion. The order of the bases A, C, T, and G varies along the length of the chromosome, and certain series of bases form the genes. Collections of genes are linked in stretches along a chromosome. All life—plants, fungi, animals, bacteria, and algae—uses this same system. Though genes determine all the properties of an organism, not all the DNA in chromosomes belongs to genes. More than 70 percent of the genetic material in human chromosomes serve other purposes, carried along as evolutionary flotsam from our ancient origins.

A complete volume of DNA containing all of the cell's genes is called a genome, a term coined by scientists to distinguish the genetic material of one species from another. The genomes of different species vary in size. The nematode (a small worm found in the interstices of plant roots) has a genome 38 times smaller than a human genome; the fruit fly's stretch of DNA is 18 times smaller; and the mouse has about the same amount of DNA. Although the sizes of genomes differ, many genes among genomes are similar. For example, one out of every four genes in the nematode is also present in humans, and the mouse and rat share over

90 percent of their genes with us. Even a rice plant has over 1,000 genes, or 11 percent, of its genome in common with humans. For decades biologists have used other animals, like the nematode, fly, and mouse, as genetic models for human biology, some of which will be addressed in upcoming chapters. The study of genes and their actions in simpler organisms implies much about human systems. Plus, science enjoys a double benefit: our knowledge of genes and genomes progresses not only because of the degree of similarity found between simple organisms and humans but because, in general, the simpler the organism, the easier it is to explore.

How many genes reside along human chromosomes? Estimates vary, but the Human Genome Project, a 10-year, government-sponsored research effort recently completed to determine the sequence of DNA in the human genome, lists about 25,000 genes. Added together, the human genome contains over 3 billion chemical letters. If all the letters in the genome of a single human stem cell were put on paper, the amount of information would fill one and a half million pages.

PROTEINS

How can genes determine all the characteristics of living things? The answer lies in the fact that each gene codes for a specific protein. Proteins determine how cells work and how we look—whether we are tall, short, dark or fair-skinned. There are many different kinds of proteins. Proteins such as collagen provide bodily structure by connecting tissues and organs. The protein hemoglobin carries oxygen in the blood. Some hormones (chemicals that regulate body functions) are proteins. For example, the hormone insulin regulates sugar levels in the blood. Another example of proteins are the workhorse enzymes that nudge—and sometimes kick—chemical reactions to completion.

Amylase is an enzyme in your saliva that helps break down food for digestion. Proteins are responsible for the characteristics of an animal's species, such as whether they walk upright or scramble on all fours, or whether they have a four-chambered heart or instead pump along with something less. Proteins run the complex machinery of the animal, and they facilitate all aspects of life. It is no understatement to say that genes and their proteins are representative of life itself and that without them you have something quite different, like, for example, a rock.

Gene expression is the term for the multistep process that deciphers a gene's information to produce a specific and unique protein. In the cell, DNA rarely leaves the nucleus, whereas proteins are found mostly in the cytoplasm. This begs a question: what transports information from the genes to the cytoplasm where proteins are made? Interestingly, an intermediary molecule called RNA is found in both the nucleus and the cytoplasm in great abundance—the concentration of RNA in any cell is up to ten times greater than the amount of DNA. It turns out that RNA is the messenger that shuttles genetic information from the nucleus to the cytoplasm.

Gene expression (shown here) begins in the nucleus, when a single strand of DNA, unwound from the helix, serves as a template for a new strand of RNA. A nuclear enzyme moves along the strand, building a chain of RNA by sequentially adding the complimentary chemical bases, but substituting a new base, uracil (U) for thymine (T).

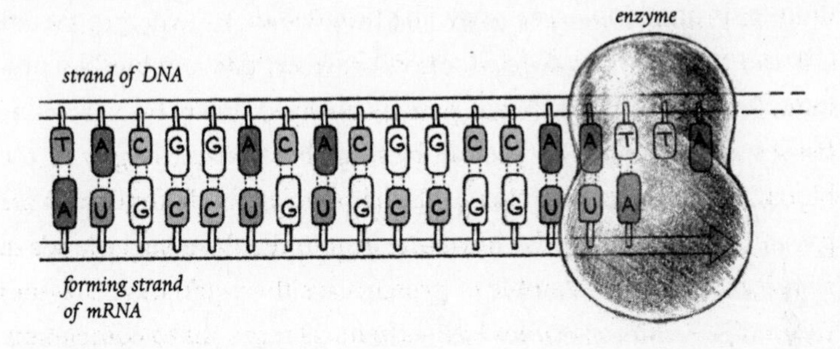

Once it reaches the end of the gene, which can be thousands of bases long, the new single strand of mRNA (the shorthand "m" for messenger) is released and moves into the cytoplasm. The end result is a version of the gene in RNA form.

Many structures "cohabitate" in the cytoplasm, most of them involved in the production, packaging, and transport of proteins. The strand of mRNA travels from the nucleus to one such structure called the ribosome, the site of protein manufacture. At the ribosome, sequences of three letters in the mRNA—AAU, CGA, UGA, and so on—are matched up with one of the twenty different kinds of protein

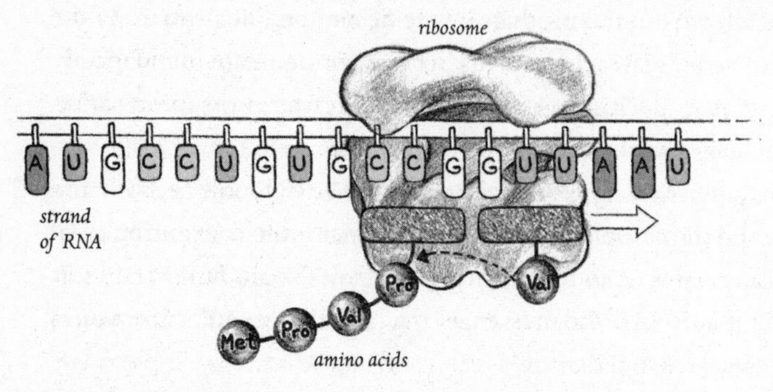

building blocks, the amino acids. For example, GUU always calls for the amino acid valine (abbreviated "Val") as shown here.

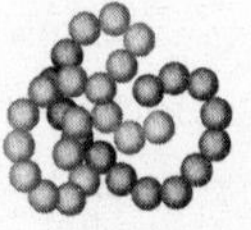
folded protein

The amino acids are linked together in long chains that fold up to form proteins, like in the illustration here, which can be thousands of amino acids long. The progression of gene expression takes information originally coded in DNA, decodes it into RNA, and then assembles into proteins, one amino acid at a time.

Gene expression, then, boils down to a few steps:

DNA → mRNA → Proteins

The three steps form a fundamental principle that is central to understanding stem cells. Though gene expression can seem complicated, the important thing to remember is that the sequence of three-base units in each gene codes for a specific protein. Even one shift in the order of bases changes the protein that is coded for and, in some cases, even renders the gene useless and the protein unavailable to the organism. As a result, the interplay between genes and proteins lies behind many, if not most, illnesses. Thousands of disorders, such as cystic fibrosis, sickle cell anemia, and Huntington's disease, result from a single defective, or mutated, gene making a dysfunctional protein. Heart disease, arthritis, cancer, and diabetes are likely caused by a combination of environmental effects and mutations in many genes. Dozens of genes spread among seven different chromosomes may be responsible for breast cancer alone.

Not only can stem cells help us understand how disease begins, they also offer potential medical therapy by replenishing or replacing cells and tissues destroyed by disease. Aberrant genes might be corrected in stem cells, then given back to the patient, resulting in a fully functional protein. Research and therapies using both approaches are explored in Chapters 6 and 7.

The interaction of genes and proteins defines the peculiar potency of stem cells and holds the secrets to how an organism grows and develops normally. For example, if all of an animal's cells contain the same DNA, why are cells different? The differences are due to gene expression: which genes are turned on, or expressed, in some situations but not in others. During normal development, many genes are expressed at once only to shut down as new sets of genes swing into action. Waves of gene expression control the changes from egg to embryo to fetus—and finally to the fully developed animal present at birth. In contrast, abnormal gene expression is responsible for developmental problems, like birth defects. Manipulating stem cells in the laboratory can actually model the steps a disease takes during its formation, allowing scientists to understand how diseases arise and to experiment with genetic and chemical therapies.

AMALGAMATED CELLS, LTD.

Cell division is a hallmark of biological growth and development, whether it is in the womb, after birth, or after an injury or insult. Genes orchestrate cell division, but they also direct an animal's development—no small feat considering that between fertilization and birth one cell must become millions of cells, and these millions must be organized so they can function properly.

Stem cells change by virtue of the interactions between their genes and proteins. The most protean variety, the embryonic stem cell, is masterful at using its genes to change into any of the hundreds of cells and tissues present in an adult organism. A distant relation, the adult stem cell, is not as powerful, but it is pivotal in maintaining the animal in good working order. As a result, both varieties of stem cells are the subjects of intense study.

Three basic features of cells are important to understanding stem cell biology. First, genetic information is contained in the nucleus, and every cell in an organism shares the same genes. Second, cells duplicate themselves by way of cell division, and in so doing pass their genetic information to new cells. Third, many iterations of gene-guided cell division must occur for an organism to fully mature, as we shall see in Chapter 3.

Life begins with a single cell. One cell, dividing into two, then two into four, and four into eight until there are billions of cells: patterned and diffuse, color-coded and clear, working-class and upper-crust, ancient and young, assembled into a great, thriving mass that is the complete organism.

3

How We Get to How We Are

We are standing and walking with parts of our body which could have been used for thinking had they developed in another part of the embryo.

EMBRYOLOGIST HANS SPEMANN

The march of a single cell to a fully functioning animal has long captured the imagination of biologists. For over a century, scientists have acknowledged that the egg holds the secrets of human development. Recently, however, some of the cells that arise from a fertilized egg have emerged as star players: the embryonic stem cells (ESCs). Formed mere days after fertilization, they are central to the deliberate change orchestrated by genes and their encoded proteins. Indeed, stem cells are responsible for every one of the more than 200 types of cells present in a fully developed human.

Human development begins with just one cell—the fertilized egg. It uses mitosis to divide rapidly, each new round of cell division guided by a complex cascade of gene action. The final result is not a simple diversity of cells and tissues, but an integrated system, a fully functioning living thing. But, human development does not stop at birth. Even after we're fully grown, we replace billions of old cells each day with new ones, we heal after we are injured, and our cells respond to and defend against environmental insult, such as bacteria or viruses.

EGGS AND SPERM

Cells pass genetic information through mitosis, creating a new cell genetically identical to the original. But during fertilization, two cells—an egg and a sperm—combine. This presents a logistical problem, genetically speaking. Imagine two garden-variety human cells with forty-six chromosomes, one from a male and one from a female, joining in a kind of faux fertilization. The resulting fusion would be a chaos of ninety-two chromosomes. If this jumble were the beginning of life, then some serious sorting would be needed: there would be four copies of each gene, a copy from the female's two parents and a copy from the male's two parents. But organisms that reproduce sexually have instead evolved a solution called meiosis, cell division that *reduces* the number of chromosomes by precisely half. The resulting egg or sperm cell, or gamete, has twenty-three chromosomes. When egg and sperm unite through fertilization, the new cell that results is restored to the full number required for normal development: forty-six. How gametes form is an interesting story—actually, two stories—because eggs and sperm follow different developmental pathways.

In human males, the formation of sperm starts at puberty and continues into old age. Spermatocytes, precursor cells that become sperm, divide through meiosis, each spermatocyte producing four sperm cells. A typical cycle of sperm formation takes about two months, each cycle forming millions of cells. By contrast, the female at puberty has a bank of thousands of oocytes, the cells that eventually become eggs. However, only one egg matures from an oocyte each month—about 400 during a female's reproductive lifetime. Meiotic cell divisions that yield four sperm in males result in only one egg in females: meiosis in oocytes creates unequal allotments of cytoplasm that result in small, nonfunctional cells that eventually degenerate and disappear.

THE COLOSSUS

The human egg is huge—250 times bigger than a normal red blood cell, and it weighs 100-fold more. Its nucleus, reminiscent of the great swirling spot of Jupiter, dominates the inner reaches of its cytoplasm. Those sperm that reach the egg's outer surface look like explorers on a vast, curving landscape. (The sheer number of male gametes make up for their small size. On average, over 100 million sperm attempt to reach the egg.) Once fertilized, the egg becomes the ancestor of all cells, including stem cells. Within it lays a nascent power that science has only begun to understand.

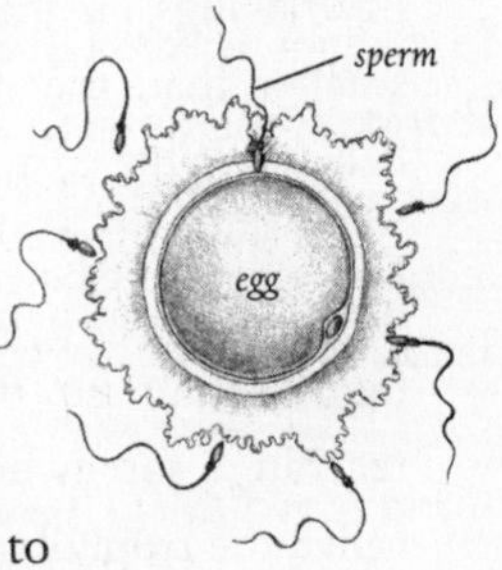

Fertilization occurs when one sperm penetrates the egg's outer membrane, as illustrated in the figure to the right. The chromosomes from each parent merge in one nucleus, and development begins. Little is known about the delicate process of early human development—much has to be inferred from studies of other animals. But animal research shows that both the male and female genomes are required to start the process. After fertilization is finished, the egg is called a zygote (from the Greek *zygous*, meaning *yoked*). At this stage the zygote—the earliest form of the embryo—is just a single cell, capable of becoming a complete organism. Survival isn't a sure bet—nearly half the zygotes die at this stage. But if it thrives, it will embark on a remarkable journey. It will become trillions of cells; its infinitesimal mass will grow to well over a hundred pounds, and its individual character will change into dozens of complex organs, such as the liver, the heart, and the brain.

THE CLARION CALL FOR STEM CELLS

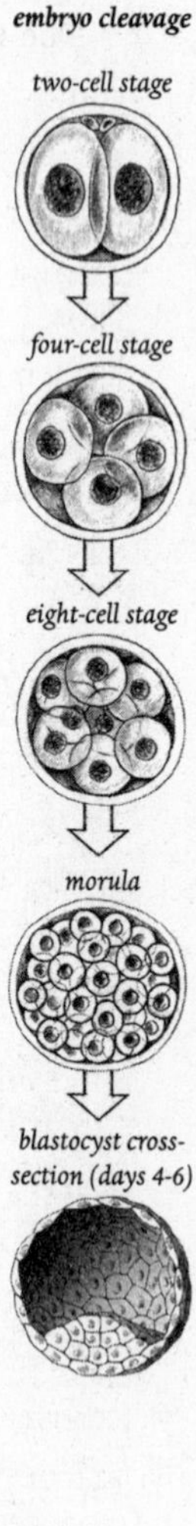

About one day after fertilization, the formation of the embryo begins. The zygote begins a series of mitotic divisions, collectively called cleavage, that result in smaller and smaller cells. Imagine dividing a perfectly plastic soap bubble (one that won't pop when you cut it) with a knife. Where there was one bubble, now there are two smaller bubbles. Cut it again, and four bubbles appear. In the zygote, a similar process repeats for several more cycles, resulting in eight cells, sixteen cells, thirty-two cells, and so on, surrounded by a transparent membrane. Just like our soap bubble, the total volume changes very little as the zygote divides. The progression of drawings on the right shows that each new cell is smaller than the one that preceded it. The interior becomes quite crowded with densely packed cells.

At three days the embryo resembles a mulberry, hence its Latin name, the morula. On the fourth day, fluid passing into the morula forms a cavity. The cells rearrange somewhat into two regions, an outer layer called the trophoblast (from the Greek *trophe*, or nutrition) and a few dozen internal cells, called the inner cell mass or ICM, the source of embryonic stem cells. At this stage, between four and six days old, the developing embryo is called a blastocyst. *Embryonic stem cells* is a

embryo cleavage

two-cell stage

four-cell stage

eight-cell stage

morula

blastocyst cross-section (days 4-6)

term used for laboratory cultures of cells made from the ICM of blastocysts. The genes involved in development have begun an important initial task: instructing the cells of the trophoblast to change into the placental and amniotic tissue that will eventually support the embryo. These genes also send cues to the ICM that eventually cause it to develop into the embryo proper.

The term *embryo* is easily misunderstood when taken out of the context of early development. The zygote, the morula, and the blastocyst are the earliest kinds of embryos, and there are nineteen other recognized stages before fetal development begins. Embryologists also call the blastocyst the *early* or *pre-implantation* embryo because it has not yet attached to the uterus. The four-day-old embryo—the blastocyst—is a corpuscle about 0.1 millimeter across, smaller than the period at the end of this sentence.

VIVE LA DIFFERENTIATION!

As the human blastocyst moves into the first week of development, the cells of the ICM begin to differentiate, or change. Differentiation is the sequence of events that lead stem cells to become any one of the hundreds of cell types in the body. Every passing hour represents a turning point for a stem cell, and the trajectory of change is overwhelmingly in one direction: cells are destined to become specialists in a fully functioning organism. Once a cell has embarked up a certain path, only the rare exception—scientists aren't sure if this happens naturally at all—will retrace its steps to become a younger version of itself. Differentiation is progressively restrictive: an ESC cannot create an acid-secreting stomach cell or a bacteria-engulfing white blood cell in one step. Instead, a cell first becomes a member of a group of cells destined for certain tissues and organs. Further steps reduce its options until it reaches its final destiny. In fact, the four-day-old blastocyst is already transfigured—the trophoblast and ICM are the earliest examples of cell differentiation. These two cell varieties have descended from a single-cell ancestor, the fertilized egg.

Genes and proteins bring differentiation about. Estimates are that more than fifty percent of the genes in vertebrate genomes play a part in development. But if all cells in the organism have the same set of genes, how then do cells develop differences? The answer lies in changes in gene activity, which in turn causes cells to change. Thousands of genes turn on and off as each gene makes proteins. Some of those proteins turn around and shut protein production down. Proteins function as messengers, traveling to other cells or dispatching other molecules to do so. They shuttle across the cytoplasm to the cell membrane, where they engage other messenger molecules in the space between cells. The messenger molecules then carry the signal to neighbors. Some proteins function solely for the internal purposes of the cell, becoming enzymes and part of the internal machinery that regulates cellular activity.

A BLUEPRINT FOR DEVELOPMENT

After the formation of the blastocyst, cell division takes on new dimensions in time and space. No longer just a ball of cells, the embryo begins to organize in three dimensions: front to back, left to right, and up and down. Stem cells begin to segregate and divide, carrying instructions that determine whether they become skin or gut, or land in some other far-flung outpost. In early development, many cells travel on one-way tickets to their final destinations.

Until this time, the embryo floats free as it journeys from the fallopian tube into the uterus. However, at about six days after fertilization, the dividing embryo confronts a crucial step in its development. It must now implant in the uterus, as shown in the diagram.

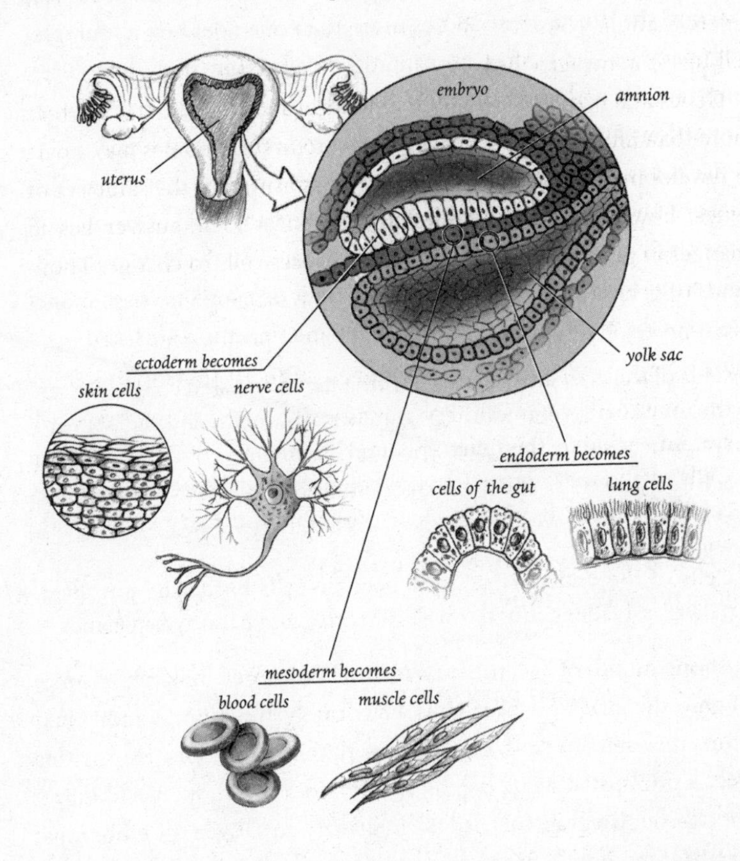

If implantation succeeds, the structures that support the embryo, such as the amnion and yolk sac, begin to form. If the hormonal cues of the mother aren't suitable, the embryo won't attach to the uterine wall. Up to half never implant and die at this stage. (Because implantation is a hit-or-miss proposition, clinics that perform *in vitro* fertilization procedures—which fertilize an egg in a test tube, incubate it for few days and then introduce it into the uterus during the cleavage stage—always have extra zygotes frozen in case implantation fails. Doctors also implant more than one embryo to increase the chances one will attach to the uterus. Sometimes more than one does and a multiple pregnancy results.) After implantation, cells from the trophoblast penetrate the endometrial lining of the womb and begin to establish connections for hormonal regulation and maternal blood supply.

At about two weeks, the embryo enters a new phase that signals the start of a body blueprint. Cells from the ICM change and organize into three kinds of cells called germ layers, illustrated on the previous page. Cells from each germ layer begin to further develop into tissues and organs of the body (only major cell types and structures are listed):

- Cells of the ectodermal layer become nerve and skin cells and form the inner ear, eye, mammary glands, nails, teeth, and the nervous system, including the spinal cord and brain.
- Cells of the mesodermal layer become blood, muscle, and bone cells, and make the heart, skeleton, gonads, urinary system, fat, and spleen.
- Cells of the endodermal layer become cells inside the gut, liver, pancreas, bladder, lungs, tonsils, pharynx, and parathyroid glands.

At one month of age, the human embryo is three millimeters long, roughly the size of a pea. Now cells rarely function as individual actors; they start to work in concert with their neighbors. During this time, a primordial heart begins to beat. In a few days, it will pump blood through a graceful triad of circulatory arcs. Even the embryonic heartbeat, without a functioning system of blood vessels, is essential to development. Recent research in zebrafish completed at the University of California, San Francisco, has found that the very action of a beating heart is a signal for the formation of proper functioning valves. Apparently, blood flow sets certain cells on the trajectory to become part of a heart valve. If true, this information could lead to the prevention of *in utero* defects caused by the interruption of the embryonic heartbeat in humans.[1]

The early days are crucial. Signals between cells increase, causing more cells to differentiate. The basic form of the body is laid down when, along with a primordial heart, the embryo develops structures that will become the future spinal cord and brain and small protuberances (called limb buds) where the limbs will be. The optic vesicle, a slight depression where the eye and optic nerves will grow, rests in a cusp along the mushroom-like swelling that will become the cerebrum. The embryonic stem cells have done much work in just a few weeks.

FROM STEM CELLS TO ORGANS

During the next four weeks, all the major organ systems appear. The embryo changes from a beanlike crescent to a form with undeniably human characteristics. At the end of two months, the embryo has recognizable limbs and a heart, along with a brain, eyes, ears, and a nose. Perhaps the most astonishing accomplishment during this phase is the formation of the nervous system. Neural anatomy is not a single organ—it is an intricate system of organs and conduits routed throughout the body. Following the fate of embryonic stem cells along one pathway in the laboratory shows how convoluted their journey can be.

The outermost germ layer, the ectoderm, eventually becomes part of the body's entire sensory and motor apparatus. An opening cascade of cell signals causes the topmost portion of the ectoderm to further diversify into specialized cells of the nervous system. Brain-related structures must be planned and executed simultaneously by other embryonic cells. For example, the brain needs a container to grow in, so genes in a different layer, the mesoderm, must coordinate the development of the skull. Other genes must coordinate the pace of growth so that the skull will develop at the same rate as the brain. As they gain mass through cell division, the brain and skull will require the support of muscles and tendons formed by different kinds of mesodermal cells.

Forming the brain itself is a monumental undertaking. The human brain is organized into layers, each containing uniquely shaped neurons with special connections to its neighbors. During the early weeks of development, groups of genes direct the first major divisions in the brain, which include the front, middle, and back sections (called the forebrain, midbrain, and hindbrain, respectively). Cells that form the backbone develop in synchrony with cells that become the spinal cord and coordinate their connections. Likewise, cells that become motor neurons change in concert with the forming appendages: each muscle, tendon, and bone will need enervation. Neurons at the outer extremities race to establish longer and longer chains of connections in order to maintain contact with the spinal cord as the embryo grows. Neural cells bordering skin cells must bring information back to the brain.

STEM CELLS BEGET STEM CELLS

Which cells are essential for nervous system development? Things begin with the inner cell mass in the embryo, which changes into the embryo's ectodermal cells. Another round of differentiation makes neural stem cells. These so-called adult stem cells—such cells have restricted developmental potential and are present throughout the creature's life—divide especially quickly. The resulting daughter cells are quite different. One functions like the parent and remains behind for another round of division. The other, called a progenitor cell, becomes the antecedent of twenty different kinds of cells present in the central nervous system.

Three major varieties of cells arising from neural stem cells are shown here: oligodendrocytes, astrocytes, and neurons. Oligodendrocytes make myelin, the material that wraps and insulates nerves and is essential for conducting nerve signals along the spinal cord. The star-shaped astrocytes, the most abundant cells in the brain, provide structural support and assist with learning and memory. The most important cell made in this pathway is the neuron itself. Degeneration of neuronal cells causes a variety of diseases and dementias, including multiple sclerosis (involves oligodendrocytes), Alzheimer's disease (involves astrocytes), Huntington's disease (involves motor neurons), and Parkinson's disease (involves motor neurons), among others. Abnormal growth of many of these cells causes various types of brain cancer. Neural cells are of interest to biologists because damage to, or death of, these cells is often permanent. Like other mature cells, when a neuron is finally made, mitosis ceases—new cells cannot be made.

Found in small numbers in the fully developed animal, adult stem cells usually reside in the organs or tissue specified by their germ layer heritage. Adult stem cells (like the neural stem cell) retain their ability throughout adulthood both to renew themselves by mitotic division and to generate numerous fully mature cells in the body. Every adult stem cell traces its origins back to the embryonic stem cells of the blastocyst.

At eight weeks of development, the work is nearly

from stem cell to neuron

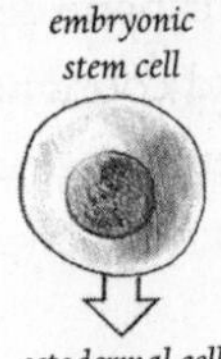

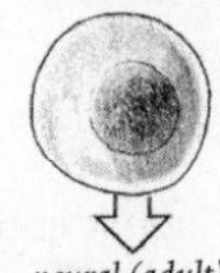

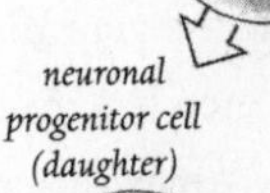

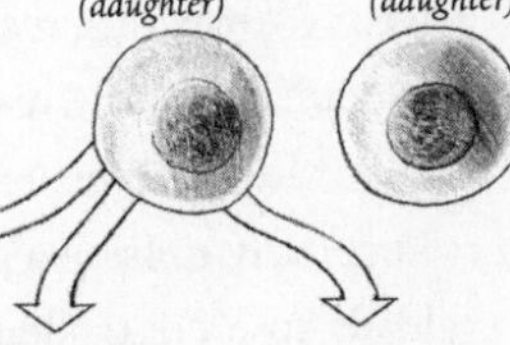

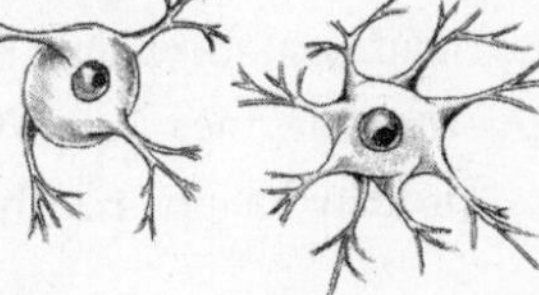

complete. All the major systems have begun to form, and the embryo enters the fetal stage. Fetal development is a period of rapid growth: adult stem cells harden bones, form internal organs, complete the skin, make blood, and establish nervous and circulatory systems. Seven months later the fetus is poised for birth. Over a billion sensory cells await their first signals from the outside environment. At birth the gray labyrinth of the brain weighs just over a pound and is packed with over 10 billion cells. It is perhaps the finest and most complicated biological structure ever assembled. And it all began with the spiral of four chemicals that make up the genes.

ANIMAL MODELS

When it comes to our understanding of human development—how a fertilized egg becomes an entire organism—we can thank many animals that are used as models to study development, including the mouse, the monkey, and the sheep. Other important model organisms include the zebrafish, the fruit fly, and the nematode.

Science has long followed *failures* of development, that is, genetic wrong turns that result in deformity and disease. Most disease has a genetic cause, and disease can be inherited just like blue eyes or brown hair. Much can be learned from wrong turns in animal development. Purposely causing a genetic error, or mutation, in a fruit fly or mouse leads to a controlled failure in development. Observing the consequences of intentional mutations helps researchers assemble a genetic "parts list," providing insight into things as essential as the production of insulin, the enzyme that controls blood sugar levels, or melanin, the protein that protects the skin against cancer.

BREAK IT AND SEE WHAT HAPPENS

The center of Stanford University's newest biomedical research building is not inside but outside. Standing in the elliptical courtyard that dominates the interior, a visitor is startled to see not so much a building as a circular expanse of azure California sky. The courtyard is encompassed by sigmoidal stretches of polished steel. A *basso profundo* thrumming from the towering air exchange columns four stories above leaves the undeniable impression that this building lives. The energy is palpable: Behind the massive curved windows and across the cantilevered bridges, people are in motion—walking together or solo, arguing, waving at a white board full of equations, shuttling back and forth between laboratory benches that crowd cafeteria-sized rooms. Hundreds of scientists work at the Clark Center—computational chemists, bioengineers, biologists, and physicians. In step with all this buzz, the collection of research programs even has a James Bond name: BIO-X.

Strolling around the Clark Center is like walking inside a man-made cell. In the nucleus, a glass and steel office sequestered along one of the inner curves, sits the center's director, bemused developmental biologist Matthew Scott. Scott is youthful and tall and has a long, full beard that grows down to the second button of his shirt. Today a necktie, the knot listing slightly to the right, covers that button and ends just above the fourth. He's meeting a politician later—it's clear that neither meetings with politicians nor neckties are among his favorite things.

He'd rather talk biology. "I've just returned from visiting a cave in Southern Arizona. It had been sealed for over three decades," he begins, eyes widening. "The microclimate kept the cave in perfect condition. It recently opened to visitors. But soon after, naturalists found something was profoundly disturbing the formations inside—things were disintegrating. Guess what caused it?" He pauses for emphasis.

"Dead skin! Humans drop millions of skin cells every day, and the growing pile was wrecking the cave! They solved the problem by vacuuming everyday."[2] Scott makes an effortless transition from geology to biology, specifically, the biology of cell renewal.

Like many developmental biologists, Scott's own research deals with model systems, which can be either *ex vivo* reproductions of biological processes or *in vivo* work in animals such as the fruit fly, nematode, or mouse. He is decidedly egalitarian about lower life forms. "Animals are good for study because each animal is different. Speech, for example, is an attribute that distinguishes humans. But other organisms have attributes that we don't. For example, birds evolved a complicated wing structure for flight, and salamanders can regenerate severed limbs. From a genetic point of view, it's not that animals are higher or lower than we are, it's just that they've adapted differently, and we can study and learn about these adaptations."

Genes and proteins aren't the only "least common denominators" between humans and other animals. For nearly a century, scientists have known that the basic biology of the one is nearly the same as the others. Our understanding of processes such as energy metabolism, the chemistry of digestion, and heredity come from our patient study of plants and animals. Today the challenge, according to Scott, is figuring out which genes are responsible for guiding an organism's development: what makes it change its shape or interact with other cells? Biologists do this is by interrupting links in the system—by breaking things genetically and then observing what happens.

Scott starts with a car. "If you are interested in a car's ability to make light, but don't know how a car works, you can break or unplug certain components, mess up the system, until finally you disrupt the thing that gives light—a wire, a battery, a bulb. The observable result is no light. Along the way, you'll identify all of the parts that are essential to a car's system of light. You can learn about any feature of the car this way, how it brakes, how it accelerates, how it uses fuel."

In order to identify the genes involved in a certain observable trait, such as what turns a normal cell into a cancer cell or a stem cell into a heart cell, Scott uses the same approach. Disrupting or mutating the genes involved alters the production of proteins. The classical approach of first introducing mutations and then observing results has been around for years. Model organisms that reproduce quickly and in great numbers, such as yeast, the fruit fly, or the mouse, give fairly quick results. Mutants in many genes along the same developmental pathway can be compared to determine which gene acts at which step.

Modern biology has transformed the old mutate-and-wait strategy. Once a gene is identified, comparing it to other known genes often help determine its function. Because genes from other animals are cataloged in databases, searching this information can lead to DNA matches that are similar to the unknown gene. Further work with that gene in a biological system can determine whether its protein acts like or unlike the known gene. There are thousands of developmental genes; each gene can have several functions, each one acting differently at different times. Advanced technology, such as silicon chips dotted with tens of thousands of bits of DNA, helps researchers determine which genes are turned on or off in a certain cell or tissue at a certain time. The Human Genome Project also serves as a parts list of human and animal genes. The challenge is figuring out what all these stretches of DNA actually do.

The vastness of this problem isn't lost on Matthew Scott. But it really doesn't matter, because the same wonder and excitement he has about dead skin is also evident in his everyday scientific thinking. He raises his hands, palms down. "Mozart used his hands to play a piano. If you boil down the ingredients of a right hand and a left hand, you'll see they are exactly the same. How did our right hand become different from our left?" It is an intriguing question, typical of a developmental biologist.

MARCHING ORDERS

Animal development requires prodigious quantities of precisely coordinated cell division. The genetic instructions delivered from the chromosomes of one cell—the zygote—begin to coordinate the formation of the embryo. The instructions are a set of marching orders to the zygote: make it this way, and this fast.

And so it does. During development, an embryonic stem cell changes both its form and its function as it divides and builds the three germ layers that will become the entire creature. In mammals, embryonic stem cells eventually change into over 200 different types of cells. Development is not just one type of cell changing to another. Cells go hither and yon, assemble in quantity, organize structurally, and work in concert.

Cell differentiation is progressively restrictive: once a cell receives instructions to become a cell of the nervous system, it cannot, in theory, become a heart cell. But the adult stem cells in nerve, muscle, or blood-forming pathways can, and do, stay behind in a primordial state in order to replenish cells and tissues. The close relations between the very potent embryonic stem cells and the not-so-potent adult stem cells are vital to our understanding of how disease, aging, and infirmity are expressed in our bodies.

4

A Brief History of Embryonic Stem Cells

The practical difficulty is in getting people to discuss things which are still hypothetical. It comes on with a bump once it becomes a reality.[1]

IAN WILMUT, THE MAN WHO CLONED DOLLY

Fifty years ago, an unfortunate laboratory mouse with a gigantic scrotum dragged himself across his cage to his water bottle. Leroy Stevens, an embryologist working at Jackson Laboratories, a genetics institute in Bar Harbor, Maine, noticed the odd gait, picked up the mouse, and turned him over.[2] A large tumor had developed on the mouse's testes.[3] Things got really interesting when Stevens dissected the mouse and sliced the tumor open; it contained a bizarre ragout of ill-formed mouse parts, including skin, teeth, bone, tangles of hair, and parts of muscle! The strange stew was a benign form of cancer called a teratoma (from the Greek term for *monster*). When Stevens teased apart the tumor and transplanted the bits into the bodies of healthy mice, they grew into a multiplicity of cell and tissue types. This rather unusual result led Stevens to suspect that the tumor, with its freakish assemblage of parts, had formed from a wayward embryonic cell gone cancerous.[4]

Stevens noticed something else about teratomas. Among the fully differentiated cells in the tumor, he also found uniform groups of smaller cells that didn't differentiate; instead, they made more copies of themselves. These rogue cells, he postulated, were the source of the trouble—not only did they make different mouse parts, they also perpetually renewed themselves. Bits of the puzzle began to fall into place. A tumor begins with a single cell; so does an embryo. If the cellular delinquent escaped from the embryo, it explained why some of his transplants grew into little cylinders of cells that looked like inside-out blastocysts. He immediately published his results: the resemblance between the cells from inside the tumor and those cells from an early mouse embryo were unmistakable. Both were very potent, both made different body parts, and the careful scripting between genes and proteins in one was catawampus in the other. The teratoma cells had all the markings of a cancerous doppelganger of the embryonic cells.

Visitors to Stevens' laboratory studied his isolation and culture techniques, obtained his strains of mice, and adapted the methods needed to grow and maintain the cells in their own laboratories. By the 1960s, other labs were growing these cancerous stem cells, called embryonal carcinoma or EC cells, and keeping them alive for extended periods of time. By regularly changing the media and dividing the new populations of cells (a procedure called splitting), the cells stayed in their primordial state and didn't differentiate. The cultures were in high demand among embryologists because mouse embryos are little bigger than a pinprick and devilishly hard to find inside a mother mouse. Once you had one, even one too hearty nudge spills its contents and ruins your experiment. What better way to study differentiation than having an unlimited source of stem cells at your fingertips? The tumor cells showed amazing flexibility. They could be induced to change into an impressive array of tissue types when different chemical cocktails were added to the media—even cardiac cells appeared, twitching with a spontaneous heartbeat. Left alone, they began to cluster and form the same hollow structures Stevens noticed a decade earlier when he transplanted them into mice. Although cancerous, the stem cells were attempting to mimic embryonic development.

Studying EC cells became *de rigueur* for nearly 20 years. Though the cells did neat tricks in the laboratory, they also had an abnormal number of chromosomes. It became clear that study of normal mammalian development needed, well, normal stem cells. Finally, a lab in England figured out a way to grow super-sized mouse blastocysts (they described them as little "zeppelins") that contain hundreds of embryonic stem cells. By 1980 the transition was complete: lines of mouse embryonic stem cells were happily dividing in the same media developed for the carcinoma cells. A series of papers from scientists at the Institute for Cancer Research in Philadelphia, Cambridge University, and the University of California, San Francisco, confirmed that mouse embryonic stem cells not only outlived their cancerous counterparts, but also had a much greater ability to differentiate into all the major cells and tissues, including egg and sperm.[5] Researchers quickly adapted the protocols developed to isolate and culture mouse embryonic stem cells for other mammalian systems. Thanks to an unlucky mouse in Bar Harbor, Maine, embryologists like James Thomson had 40 years of research to rely on as they went about perfecting the methods used to grow first primate and then human embryonic stem cells.[6]

EMPTY EGG SEEKS NUCLEUS FOR LTR

About the time Leroy Stevens began to probe the testicles of his cancerous mouse, developmental biologists were asking some far-reaching questions about the mysteries of cell differentiation. After all, teratomas are spectacular failures of development, an upset in the precise pathways needed to organize cells into tissues, tissues into organs, and the whole thing into a functioning creature. Development is a pattern of gene expression, so the chaos of a teratoma might be due to genes expressing themselves at inappropriate times. Indeed, a cancer researcher working with Stevens inserted a teratoma cell into an early stage mouse embryo, and something about the environment inside the embryo caused the cell to lose its cancerous qualities! If development was just a matter of genes turning on and off, could an embryo change the gene expression of other cells? Could an older, differentiated cell have its genes reset to an earlier, younger version of itself by being put into an embryo?

Robert Briggs, a cancer researcher working in Philadelphia, and his collaborator, Thomas King, a Columbia University embryologist, began to tackle these questions in the 1950s and 1960s. The duo focused on the large, plentiful eggs of *Rana pipiens,* the leopard frog. The diagram on the next page shows how Briggs and King removed the nucleus from an unfertilized egg cell by sucking it out with a very fine hollow needle called a micropipette.[7] In the same fashion, they removed the nucleus from a cell inside a developing frog embryo. Injecting the nucleus into the empty egg tricked it into thinking it was fertilized, and once it was fooled, the process of embryogenesis began. A tadpole hatched and grew into a vigorous, healthy frog. This was the earliest version of nuclear transfer, where a nucleus without a cell is inserted into a cell without a nucleus.

Three consequences of the Briggs and King experiment are worth noting. First, they made an animal *without* sexual reproduction—*no* sperm was involved and hence *no* fertilization occurred. Second, a genetically identical frog developed from the nucleus—a clone of the original frog embryo. Third, they transferred nuclei of different

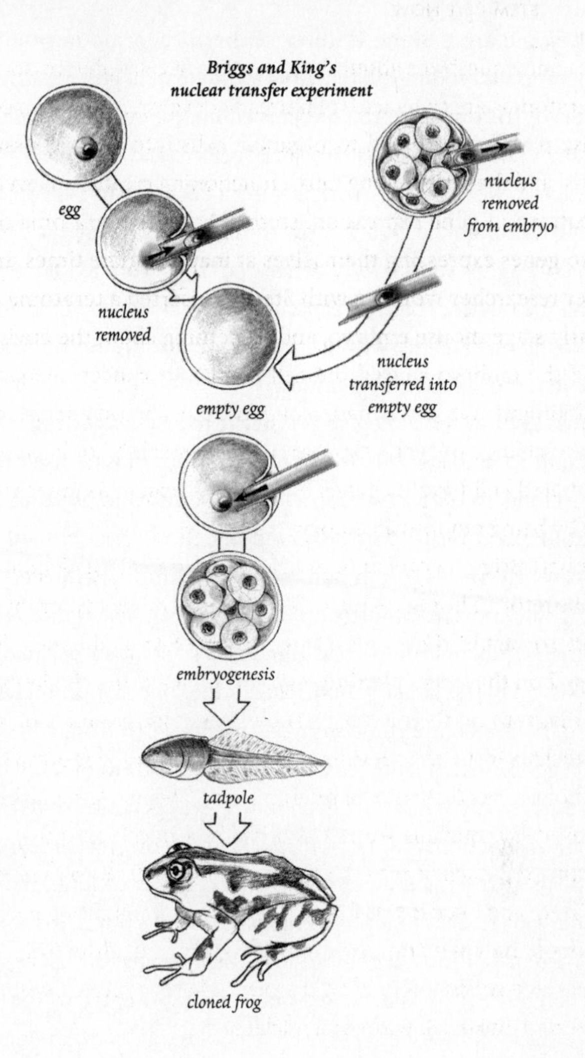
Briggs and King's
nuclear transfer experiment
egg
nucleus
removed
nucleus
removed
from embryo
empty egg
nucleus
transferred into
empty egg
embryogenesis
tadpole
cloned frog

embryonic ages. Some of these embryos became normal tadpoles. However, succeeding experiments found that, in general, the older the embryo from whence the nucleus came, the more difficult it was to obtain a normal frog. The nuclei of cells found in adult animals, they predicted, would never work in such a fashion.[8] In 1975, the impossible proved possible. Working at Cambridge University, John Gurdon and colleagues successfully generated frogs from a differentiated cell using the nuclear transfer technique.[9] The evidence of the egg's power to change a nucleus was an important result, and nuclear transfer research moved to other animals.

Just as stem cell biologists found the teensy, finicky mouse embryos difficult to manage, others were frustrated trying to use them to create a clone. In 1981, however, a German embryologist, Karl Illmensee, announced a result that beggared belief: the first cloned mammal. Apparently, the German had refined the nuclear transfer technique using fine movement instruments called micromanipulators and vanishingly thin micropipettes. Illmensee's virtuoso performance was published in the January 1981 issue of the journal *Cell.* Like Briggs and King, he used embryonic stem cell nuclei taken from a blastocyst. Of 363 eggs receiving a nucleus, he reported the birth of three live mice, a 0.8 percent success rate.[10] Each mouse was genetically identical to the embryo that donated the nucleus. Had the original embryo survived to birth, it and the new mouse would have been clones of each other. Other labs rushed to repeat Illmensee's results but failed to do so. The *Cell* paper fell into disrepute. Scientists began to believe that mammalian nuclear transfer would never work. After trying for three years to replicate Illmensee's work, Davor Solter, a developmental biologist working at Philadelphia's Wistar Institute, soberly wrote in a 1984 issue of *Science,* "the cloning of mammals by simple nuclear transfer is biologically impossible."[11] It turns out that Solter's indictment of mammal cloning was premature. The problem was the ubiquitous laboratory mouse.

Steen Willadsen was a Dane and veterinarian who worked at the British Agricultural Research Council's Unit on Reproductive Physiology and Biochemistry in the 1980s. His expertise was livestock embryology, and the government research facility, jammed with sheep, was a perfect place to try nuclear transfer. For starters, ovine eggs are larger and easier to work with than mouse eggs. Rather than using Illmensee's pipette method to insert the nucleus, Willadsen tried a less invasive method. Introducing a protein that melts the cell membrane, he fused the nucleus from an eight-cell sheep embryo with an unfertilized egg whose nucleus had been removed. He tried the method on three eggs and implanted them into surrogate ewes. Five months later, a lamb was born—a stunning 33 percent success rate. The unnamed progeny of his 1984 protocol is recognized as the first mammalian clone.[12] The result made barely a ripple in the mainstream press, but it paved the way for the big news about Dolly over a decade later. Though Willadsen had surmounted the barrier to cloned farm animals, mice continued their stubborn resistance to cloning. Fifteen years would pass before the first cloned mouse was born in a laboratory in Hawaii.

DOLLY

One year before James Thomson's human embryonic stem cell story broke, the wires fairly crackled with news that Ian Wilmut, a Scottish animal researcher, had cloned a sheep. That sheep, of course, was Dolly. Dolly made the cover of the journal *Nature* only partly because she was a mammalian clone—Willadsen's lamb had, after all, been born a decade earlier. The big news was the type of donor cell used to create Dolly; it came from a somatic cell (from the Greek word *soma*, or *body*) taken from the udder (hence the unfortunate reference to the country-western singer Dolly Parton) of a six-year-old, dun-colored,

Finn-Dorsett sheep. Until then, the effectiveness of nuclear transfer was dependent on the developmental distance between the developing egg and the incoming nucleus. Until Dolly, animal cloning success had been restricted to nuclei taken directly from embryos. The decade-earlier assertion that mammalian nuclear transfer could never work was destined for the infamous dustheap of "never coulds": Wilmut's clone came from a fully differentiated cell. Under a fetching photograph of the long-eyelashed lamb, *Nature's* headline read: *A Flock of Clones.*[13]

How did Wilmut do it? His first task: perfect the conditions needed to reset the genetic information contained in the somatic cell nucleus. To achieve this, he grew the Finn-Dorsett udder cells in a culture medium with only the minimum nutrients for survival; this arrested cell division. As the figure on the next page shows, Wilmut also removed the nucleus from another sheep's unfertilized egg. He fused the quiescent udder cell to the empty egg with a gentle pulse of electricity. After the cells fused, the proteins in the egg's cytoplasm began to reprogram the genes in the donor nucleus, resetting the developmental instructions contained in the DNA. The genes switched from their "mammary cell program" to a program that coordinated the development of an embryo. From 277 cell fusions, 29 normal-looking blastocysts developed. He implanted the tiny embryos into surrogate mothers, but Dolly was the only successful pregnancy, born five months later in July 1996.[14] Dolly was the clone of the donor Finn-Dorsett sheep. Wilmut's technique was dubbed SCNT, shorthand for somatic cell nuclear transfer.

THE TROUBLE WITH CLONED ANIMALS

Rudolf Jaenisch, professor at MIT and a stem cell expert who studies nuclear transfer, states flatly, "You cannot get normal adult animals from clones—it is not a technical problem, it is a genetic problem."[15]

Wilmut's sheep-cloning experiment

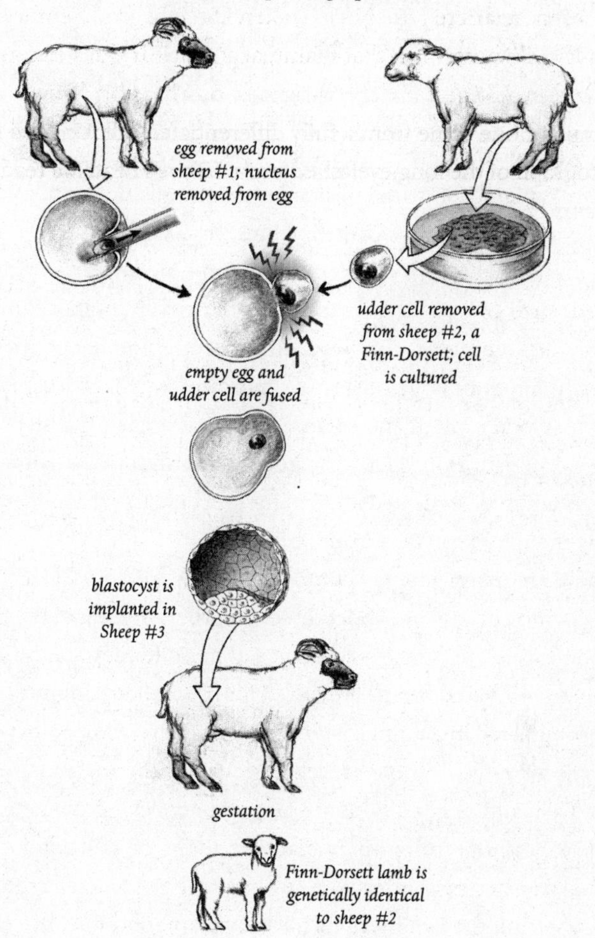

The low success rate of cloning animals may be one reason. By one count, out of 17,500 attempts at reproductive cloning in at least five mammalian species, 99.2 percent of the implanted embryos died *in utero*. Of those mammals that were born, many died soon after.[16] Cloning efficiency, the number of live offspring expressed as a percentage of nuclear transfers, ranges from 0.1 percent in pigs to 5.8 percent in mice.[17] Of the few clones that do survive, many are less than healthy. Skeletal abnormalities and arthritis are common, as are placental abnormalities, malformed organs, circulatory disorders, umbilical hernias, respiratory problems, and immune system dysfunction. Cloned mice often suffer from both abnormally high or low birth weight, depending on the procedure used. Some sheep and cattle clones suffer from large offspring syndrome (LOS), characterized by abnormal fetal growth and enlarged internal organs.[18] The Food and Drug Administration (FDA) has not yet approved meat from cloned animals and has requested all cloning breeders to refrain from introducing them for human consumption.[19]

There are many possible culprits for these problems, and the chief suspect among them is the incomplete reprogramming of the genes in the donor nucleus. Once a somatic cell has differentiated, the egg is unable to reset all the genes required for normal development. Genes can be impacted in a number of other ways. Removing embryos takes the cells away from the extracellular cues of the maternal environment. Cell culture conditions are far from perfect—growing embryos in the laboratory prior to implanting them or culturing cells for nuclear transfer may introduce fatal or troublesome changes in the DNA. Getting the timing right for transplantation into the uterus is tricky, too—if the hormonal conditions of the surrogate mother aren't spot on, the embryo doesn't implant correctly.

Perhaps other developmental failures are mechanically induced. Freezing and storing embryos and then thawing them later is not exactly gentle on the cell's biochemistry. Given the egg's fragility and tiny size, transplanting a nucleus into it is blunt and invasive. A micropipette can remove the chromosomes of an egg effectively, but it can take with it cytoplasmic apparatus essential to cell division.[20] Conversely, other cytoplasmic elements and structures may travel with the donor nucleus and end up in the recipient egg's cytoplasm, wreaking havoc by duplicating cellular machinery. Different techniques may have an impact, too. Transferring a nucleus by cell fusion is different from nuclear injection in one important way. The cytoplasms of each cell mingle during fusion, so a clone made this way has different properties than one made by nuclear transfer.

Yet, despite the low efficiencies, animal cloning continues. Using cloned animals to model human disease is an intriguing idea. Some claim that cloning a close human relative, like monkeys, for use as subjects in clinical trials would take some of the risk out of the early stages of testing drugs in humans. But though simian clones have been produced using embryonic cells, somatic cell nuclear transfer has not been successful, probably because of reprogramming effects and missing cytoplasmic factors.[21] In one experiment that produced 716 rhesus monkey embryo clones, many contained twice the normal number of chromosomes, some had odd combinations, and others were missing some chromosomes entirely.[22]

Dolly and her fellow clones went on to live reasonably healthy lives. Dolly bred successfully twice and gave birth to four lambs. She died in late middle age at age six, from a virally induced lung tumor. It is difficult to tell if her life was shortened because she was a clone. Only time and more research will answer this question.

IT DEPENDS ON WHAT THE MEANING OF CLONE IS

The twin trajectories of somatic cell nuclear transfer (SCNT) and embryonic stem cell (ESC) research collided bigtime in 1998. Among most biologists, the excitement about nuclear transfer was only partly due to Dolly. What really got them going was the egg having its way with a somatic cell nucleus—and then marching off to (apparent) normal development. Then came James Thomson, building on the work of Solter, Stewart, and those scientists one short step away, including Roy Stevens; The Institute for Cancer Research's Beatrice Mintz; developmental biologist Gail Martin from the University of California, San Francisco; and Martin Evans, Cardiff University's cell differentiation pioneer. These and other biologists perfected the nuances of embryonic stem cell culture—first in mice, then in ungulates and monkeys, and finally, in immortal lines of human embryonic stem cells. The intersection of the two technologies raised the stunning possibility that nuclear transfer could be used to make a custom-matched line of embryonic stem cells—in essence, a reserve of cells cloned from anyone who needed them to replace a diseased or injured part of their body. Wilmut showed that he could clone a viable embryo using SCNT. Thomson proved that the inner cell mass (ICM) of a human embryo could be used to make a line of long-lived embryonic stem cells. All that remained to be tried was a set of experiments bridging the two discoveries. SCNT would be used to grow a human blastocyst. Then the ICM would be removed to make a line of embryonic stem cells. In essence, an early human embryo would need to be made—then destroyed—to obtain the cells inside.

Of course, once the moral and medical implications of such an experiment became clear, all hell broke loose.

It began with a confusion of terms. When the media used the term *human embryo*, most people, including a surprising number of science and medical professionals, did not know which stage of embryo the stem cell biologists were talking about. Was the embryo recognizably human? Would removing the ICM from the embryo harm it? Would cloned humans be made in a laboratory and then be used for spare parts? Then there was the hype surrounding the promise of stem cells. The talk of curing disease overshadowed the enormous distance science had to travel to get there. People overlooked the fact that it took decades to perfect a single recipe for growing mouse embryonic stem cells—how long would it take to produce, say, a reliable source of human liver cells? The excitement about quick cures eclipsed other important reasons for using human ESCs, many of them connected to answering basic questions of biology, such as how cells differentiate and how disease begins.

Things became even more muddled. The technology involved does not fit the 30-second sound bite very well. Cloning animals and cloning stem cells share a common technique, nuclear transfer. Using SCNT blurred—in the public's mind at least—the distinction between making an animal like Dolly (reproductive cloning) and making embryos from which stem cells are derived for medical uses (therapeutic cloning). What happens to the blastocyst in these two procedures is the distinguishing factor—the blastocyst made for research or therapeutic purposes is *never* put in a mother because it is destroyed in order to use the ICM. The belief that therapeutic cloning means making humans—rather than cells—continues to be a common misunderstanding about the use of the technology. Also, *therapeutic cloning* isn't an entirely accurate term, either. No therapies have actually resulted from SCNT, and although the cells contain the same genetic information as the donor nucleus, proteins in the egg's cytoplasm may change the gene expression patterns of the embryo. The result isn't a perfect genetic clone.[23] When President George W. Bush announced in 2001 that he would restrict research to an official list of human embryonic stem cell lines, most people didn't understand what "a line of cells" actually was. So, before we move on to the next chapter, it's worth sorting out these terms and techniques, starting with the embryonic stem cell line.

HUMAN EMBRYONIC STEM CELL LINES

Human embryonic stem cell lines arise from a single dividing cell, multiplying so that every cell is like every other—a colony of cloned human embryonic stem cells (hESCs). There are two ways to produce an hESC line. One uses embryos donated from (IVF) clinics, and the other uses embryos made by somatic cell nuclear transfer. Both involve four- to six-day-old embryos at the blastocyst stage. First, consider lines made from frozen embryos donated from *in vitro* fertilization clinics—essentially the procedure that James Thomson and others use. Cleavage-stage embryos (about two days old) are thawed and put in culture where they develop into blastocysts. The outer layer of the blastocyst is removed, exposing the ICM. The cells of the ICM are removed and put into a culture dish, as shown in this diagram.

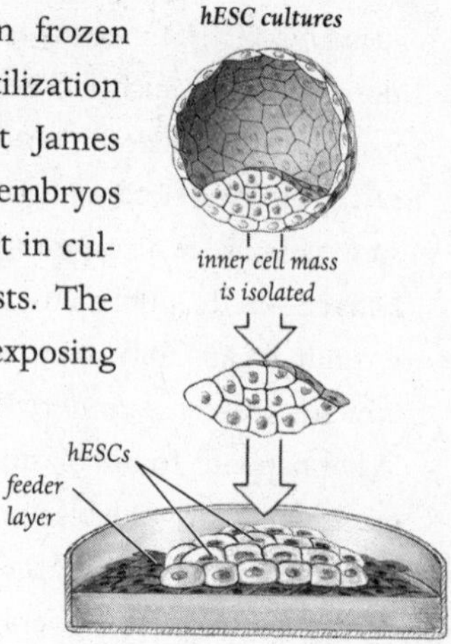

Left alone, hESCs will differentiate. To remain in their primordial state, they must grow next to a feeder layer, a densely packed mat of mouse cells stuck to the bottom of a culture dish. The feeder cells have been irradiated so that they stop dividing. Proteins secreted by the layer nourish the stem cells above, providing essential nutrients and signals that dampen the urge to differentiate. Other ingredients such as nutrients to promote growth are also added. The basic recipe for maintaining hESCs can include 16 or more separate solutions.[24] When the conditions are jiggered just the right way, a cell line is considered *established*. In this state, it can live indefinitely and proliferate, and it has stable numbers of chromosomes. Established cell lines are valuable commodities among scientists because experiments can be compared across both time and many laboratories.[25] Lines growing on mouse feeder layers could not be put into humans because the cells would likely cause an immune reaction in the patient. As a consequence, scientists have recently developed hESC lines that grow on feeder layers made with human cells or in a medium that doesn't require feeder cells at all.[26]

Using donated embryos as a sole source of hESC lines has a downside. Embryos from IVF clinics represent a very narrow genetic range: most parents frequenting the clinics are white, infertile, but otherwise healthy. Finding enough frozen embryos is also a concern. Even during routine IVF procedures, many embryos die after being thawed, not all survive the culture methods, and many never implant in the uterus. So extras are needed for subsequent procedures. If embryos used for research do survive to the blastocyst stage, some fail to make a thriving cell line. The biggest drawback is that even if thousands of lines are made from donated embryos, none of them will be an exact genetic match for a sick or injured patient. Using any such line for tissue or cell transplants will require the administration of drugs to ward off immune rejection, the same way rejection is kept at bay after an organ transplant.

HUMAN CELLS FOR THERAPIES

Dolly's unveiling signaled solutions for the scarcity of donated IVF embryos and lack of genetic diversity in the hESC lines. The biggest excitement swirled around the potential for an ultimate solution for personal medicine: an unlimited supply of replacement parts cloned from the patient's own cells. Enter another method for making stem cell lines: somatic cell nuclear transfer (SCNT). In the rush to put a name to the idea, SCNT used for medical purposes quickly became known as therapeutic cloning. As shown in the hypothetical procedure

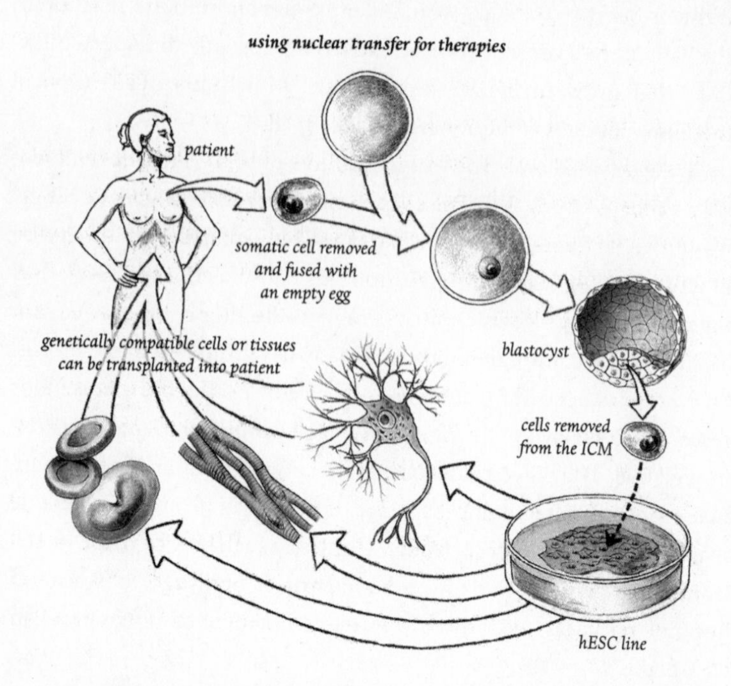

here, a cell is removed from a skin or muscle cell of the patient. As in the first steps of animal cloning, the somatic cell is fused to an empty human egg. Cell division begins, and, over the course of a few days, the zygote forms a blastocyst. To make an hESC line, the ICM is removed from the embryo and grows into a colony of cloned cells. Because all the cells contain a faithful copy of the patient's DNA, the hESC line is a good, but not perfect genetic match. Small amounts of donor DNA will be present in the egg's cytoplasm. In theory, human embryonic stem cell lines made by nuclear transfer could make healthy tissue to replace the patient's damaged or diseased tissue, which then could be transplanted. Taken alone, nuclear transfer is not a complete solution. Some drugs may be needed to fight off rejection of foreign DNA carried along with the egg. If cells die because of a genetic defect, then new cells made from the same hESC line will die, too. The trick in this case is to replace the gene with a healthy version, and then use the new line to cure the disease. Promising new ideas that feature combinations of gene and stem cell therapy are examined in Chapters 6 and 7.

The differences in the outcomes of reproductive cloning and making embryonic stem cell lines are vast. Reproductive cloning produces an animal clone. An hESC line makes cells for potential medical use and nothing more. The similarities between the two lies in the technique—SCNT—that makes the blastocyst. An hESC line cannot produce a baby because the cells are no longer part of an embryo. An embryo in a culture dish will not produce a baby either, because it needs the environment of the uterus to survive. Could a human clone be produced using the same method that created Dolly? Theoretically, yes. But because of the many biological problems associated with animal cloning, any attempt to create a human would amount to the worst form of human experimentation. As a result, every national and international scientific body vociferously condemns the idea of cloned humans. Other ethical reasons compel us to prohibit human cloning; these are discussed in Chapters 8 and 9.

The word *clone* has many meanings. Think *clones* in science-fiction movies, and we immediately conjure up vacant-eyed automatons and rapacious dinosaurs. In reality, cloning can be as simple as cell division, where daughter cells are clones of their predecessors. Most cloning happens naturally. When the fertilized egg first divides, occasionally, each daughter cell goes on to form separate embryos. The result? Identical twins, each one the clone of the other (though not a clone of either parent). Wee beasts that reproduce asexually, such as aphids, brine shrimp, yeast, and bacteria are clones of one another. Plant biologists use the term to describe a form of propagation where one plant is cut into pieces that are used to grow hundreds or thousands of identical seedlings. Colonies of cloned bacteria can produce biotech drugs or viruses grown for human vaccines.

Despite the claims of Hollywood, tabloids, and the occasional unhinged person, no human has ever been cloned, and only a few hundred warm-blooded animals qualify for this designation. *Clone* the noun and *clone* the verb are terms of biological art, part of every genetics and biology textbook. Cloning technologies are essential tools of modern life. Without them, modern genetics would still be the stuff of science-fiction novels, and we would be without scores of important drugs and newly developed therapies.

ANIMAL ENGINEERING

It's not surprising that one of the earliest uses of embryonic stem cells involved the good old white mouse. It began at the nexus of genetic engineering and embryology. Swapping genes between lower organisms, like bacteria and viruses, was commonplace by the 1970s and it wasn't long before worms, fruit flies, and mice entered the scene. Because sequences of DNA have a fair degree of correlation up and down the evolutionary ladder, scientists, ever the explorers, inserted bacterial genes into plants and swapped animal genes into bacteria and human genes into animals—the most important example being the transgenic mouse. By introducing a foreign gene into the nucleus of a mouse embryonic stem cell and then putting the cell back into a mouse blastocyst, the mouse that develops from the embryo shows the effects associated with the new genes and passes them on to its offspring. Before long, an inbred line of mice is produced—similar to a line of stem cells, but now in whole-animal form. If a gene for a human disease can be suitably expressed in a mouse, then the transgenic mouse becomes a powerful vehicle for studying disease. Cancer, AIDS, asthma, Huntington's, diabetes, and Parkinson's are but a few of the dozens of maladies studied using mouse facsimiles.

The first mouse model for Alzheimer's disease was reported in 1996.[27] Two mutations in a gene implicated in Alzheimer's had been found in a large Swedish family. Mice containing both of the genes were made and, as they grew older, showed many of the symptoms of the human disease, including the signature plaque formation in the brain and behavioral deficiencies, such as memory loss. New lines of mice with other human genes implicated in Alzheimer's are now being used to uncover the subtleties of disease progression.[28]

Studying human genes in mice has revolutionized the study of disease, but the mouse is not a perfect match for this task. Human genes often behave differently under the control of a different species, no matter how closely related. Using lines of hESCs to mimic human disorders bridges this gap, leading to faster discovery of drugs and therapies. The strategy is similar: use the hESC lines to identify events during human development, specifically, how undifferentiated stem cells become differentiated. For example, we know that cancer and birth defects are due to abnormal cell division and differentiation. A better understanding of the genetic controls of these processes would yield important information about how such diseases arise and may also suggest new strategies for therapy. Human stem cells could be used to discover new drugs. New medications could be tested for safety on differentiated cells generated from hESC lines. Cancer cell lines, for example, are used to screen potential antitumor drugs. Just like Matthew Scott's car strategy mentioned in Chapter 3, intentionally causing mutations in different genes in a biochemical pathway of healthy embryonic cells can mimic a specific disease. Knowing which genes are responsible takes us that much closer to drugs that counteract the defects.

Although it is too early to predict the economic impact of cloning superior livestock for human consumption, transgenic animals are already used to produce biological drugs such as human insulin. Herds of cloned animals containing a human gene would serve as biofactories to manufacture important human proteins that are too difficult or costly to make in the laboratory. Take human serum albumin, which is used to treat burns and to replenish body fluids lost during surgery. Expressing the albumin gene in a cow's mammary gland could produce large quantities of the protein in the cow's milk. After a purification step, the protein could be used for human use, thereby lessening the need for serum donation. Similarly, bulk production of major classes of human antibodies for therapeutic and diagnostic purposes could be achieved with sheep carrying the necessary human genes.

THE NEXT MOST POWERFUL CELL

Examining the inscrutable mysteries of the embryo led to much of our current understanding of stem cell biology, including nuclear transfer and making stem cell lines. Much has been made of their promise. But not unlike thoroughbred horses, hESCs produce an astounding variety of powerful offspring, including the somatic or adult stem cells. Biologists are still learning what adult stem cells can do and how to find them. Some varieties are rare, existing ephemerally in the organs they serve. But adult stem cells have telltale qualities, and the next chapter explores how scientists use these characteristics to identify the reclusive cells in their native environments. Any stem cell hunter worth his or her salt must have a field guide to help spot them. As we'll see, once spotted, the work is just begun.

5

Hunting Adult Stem Cells

Some circumstantial evidence is very strong,
as when you find a trout in the milk.
HENRY DAVID THOREAU

For the most part, identifying a human embryonic stem cell (hESC) is fairly straightforward. A blastocyst in hand means a ready source of hESCs. But identifying adult stem cells is another matter. Adult stem cells comprise a large and diverse category of cells destined to change from one kind of cell to another. They reside in our bodies from head to toe. In fact, as an hESC transitions to an adult stem cell and then to its final form, it typically assumes many ephemeral intermediates. Such prestidigitation gives adult stem cells a now-you-see-it-now-you-don't quality that is maddening to scientists—which leads to the central question of this chapter: what exactly *is* a stem cell?

Ask a scientist this question and he or she will most likely waffle. Even the experts have a hard time defining what is—and what is not—an adult stem cell. The major clues start with a handful of basic concepts. Stem cell biologists often mention a cell's "stemness," the idiosyncrasies that make it stand out from a crowd of millions. Two major characteristics signify a cell's stemness. The first is the degree to which a stem cell can differentiate into various cell types, commonly called its potency. The second is a special kind of cell division that ensures a future supply of stem cells.

POTENCY: 100 PROOF?

Stem cells form a hierarchy of potency. The fertilized egg and the cells that accumulate during the egg's first few mitotic divisions are considered totipotent. Totipotent stem cells are the most powerful and have the astonishing ability to become any cell or tissue in the body.[1] For example, identical twins arise after a fertilized egg divides in two. Instead of continuing to divide as part of one zygote, each totipotent cell develops into a separate animal, one a genetic copy of the other. (Nonidentical, or fraternal, twins develop from two separately fertilized eggs.) When a zygote is just a few days old, the cells inside begin to receive instructions to differentiate further and so lose their totipotence. At this stage, some cells are instructed to become the embryo, others to become the tissues of the placenta. Another cascade of signals form the inner cell mass, yet another, the three germ layers of the blastocyst. Along the way, cells become more and more specialized until they can no longer change at all.

Pluripotent cells can become most, but not all, cell types. Cells of the ICM and embryonic stem cells fall into this category. Once cells commit to a pathway of differentiation, their choices become ever more restricted, and they are called multipotent. The neuronal stem cell introduced in Chapter 3 and shown on page 33 is an example of how the fate of a multipotent cell becomes more and more restricted, eventually reaching a point where it can only become a specialized cell like an oligodendrocyte (cell that forms myelin around nerve axons) or an astrocyte (cell that gives the brain's gray matter its structure and support).

Some multipotent cells can make many different cell types. Others, such as the stem cells in the gonads, produce only one kind of cell, the gametes (such cells are sometimes called unipotent). Once a stem cell reaches the end of its pathway of differentiation, it is no longer capable of dividing—it is terminally differentiated. Skin, intestine, and blood cells are examples of cells at the end of their pathway of change. They grow old and die and are replaced by new terminally differentiated cells. Totipotence, pluripotence, and multipotence: these are fundamental characteristics of different types of stem cells.

SELF-RENEWAL

Stem cells exhibit a peculiar kind of cell division. In normal cell division, many rounds of mitosis—each cycle producing identical looking and acting daughter cells—are needed to grow a creature to adulthood and to replenish dead and dying cells during its lifetime. When it comes to stem cells, the "exact copy" rule of mitotic division changes in order to accommodate specialization. If *exact* copies were the rule, no stem cell could change into any other. That said, *part* of the exact copy rule must be maintained because without it the stem cell population would be eventually exhausted. Because of the obvious fact that stem cells are found in the developed creature, a hybrid type of cell division must exist, one that replenishes the population of stem cells even as it generates a daughter cell with more restricted developmental capacity and more specialized abilities.

Asymmetric cell division is the term for this mechanism. Symmetric cell division does a good job explaining the regular form of mitosis, when the daughter cells are exact replicas, genetically and physically. The diagram on page 33 shows how asymmetric cell division of a neural stem cell causes the daughter cells to be born *different* from each other. One cell retains its "stemness" and becomes part of the new generation of stem cells; the other cell becomes a different cell type, traveling along the pathway to specialization. In this case, the specialized, fully differentiated cells at the end of the pathway are oligodendrocytes, astrocytes, and neurons. The genetic complement of each cell remains the same. Asymmetry, then, is where the stem cell either divides in order to restock and renew, or changes to a more mature and restricted cell type. Self-renewal and asymmetry are two hallmarks of stem cell behavior.

EXCEPTIONS TO THE RULES

One sobering lesson scientists learn early in their training is that the neat and tidy frameworks that define their discipline are soon shot through with exceptions. The rule "specialization goes one way" can be broken in the laboratory—adult cells can be enticed to dedifferentiate, or go into reverse gear. The best example of man-made dedifferentiation is somatic cell nuclear transfer (SCNT), where a terminally differentiated nucleus is reprogrammed to make an embryo. The fact that SCNT works makes sense from a genetic standpoint because genes don't disappear during development; they are simply expressed at different times and in different combinations.[2] In theory, if genes are turned off during the changeover, it is possible to force things in the opposite direction developmentally by turning the genes back on again.

Can an older, stodgy, nonstem cell find an elixir that turns it into a handsome, vibrant stem cell? Biology says perhaps. There is no conclusive *in vivo* evidence to suggest that cells revert developmentally in the adult animal, but genetic reprogramming may work in the laboratory. If scientists could find this reprogramming in action, they might be able to use what they learn to develop a reliable way to reprogram older cells to their earlier versions without using nuclear transfer—or eggs and embryos. Reversing cells one step at a time in a culture dish could show which genes control the onset of disease and when, and the technique could be used to make many identical copies of adult stem cells for therapies. Chapters 6 and 7 examine reprogramming and dedifferentiation in more detail.

Other rules of cell fate may also have exceptions. Recall that once an embryonic or adult stem cell gets its instructions to change, it stays on a specific pathway: a cell first becomes either endoderm, mesoderm, or ectoderm, and then on to the successive cell type along a pathway specified for individual tissues and organs. But experimental conditions might disrupt these restrictions. For example, it is possible to fuse a human liver cell with a mouse muscle cell, exposing the human nucleus and its DNA to proteins in the mouse cytoplasm. This strange hybrid turns off the human liver genes but turns on a different set of genes that make human muscle proteins. Recipes of chemicals can switch genes on or off. Neural stem cells placed into culture dishes with just the right combination of ingredients can produce insulin.[3]

In recent human clinical tests, stem cells from a bone marrow donor were found, surprisingly, not only in the recipient's blood but also in tissues and organs. A flurry of animal research results in the late 1990s hinted that blood stem cells are pluripotent—able to change into neural, muscle, skeletal, liver, kidney, lung, and skin cells, types not normally associated with their original germ layer. Other studies claim that cells from the brain can change into blood cells. These results remain controversial. Once a cell is committed to its fate, the textbooks say, there is no turning back and no switching from one path to the next.

If a stem cell jumps from the germ layer pathway where it began and lands in a pathway of another germ layer—or switches from one kind of cell to another, acquiring new functional and physical characteristics—the change is called transdifferentiation. Plasticity describes cells that have the ability to transdifferentiate or dedifferentiate. When first reported, plasticity was hailed as a significant departure from biological dogma. It also suggested that the solution might lie in using adult stem cells for therapeutic use, circumventing the ethical worries associated with embryonic stem cells. In this hopeful scenario, plentiful blood or skin stem cells could be changed into colonies of rare pancreas or liver stem cells for transplants and then banked for future use. But proof that true adult stem cell plasticity remains controversial. The evidence both for and against plasticity is continually argued, disputed, and challenged in journals and scientific meetings.

THE PROGENITORS

Stem cells often make intermediate transitory cells. Just as physicists continue to discover new subatomic particles, biologists are uncovering a host of transitory cells, often described as progenitor cells. Progenitors multiply rapidly by mitosis, adding critical numbers to a depleted population of elderly or dying cells. Like true stem cells, progenitors can be multipotent, forming disparate sorts of cells. Other distinctions between stem cells and progenitor cells are not always so clear. In both humans and mice, adult stem cells can usually divide indefinitely, but human progenitor cells divide only a limited number of times. What's more, mouse progenitor cells can be coaxed to become true stem cells by adding the right chemicals to the culture dish.[4] Controversial evidence suggests that human progenitor cells can revert to ancestral types of cells in both the intestine and the brain. In both instances, the effect is caused by injury to the tissue where the cells reside.[5]

Not surprisingly, progenitor cells are objects of intense research. For organs and tissues that don't naturally repair easily, such as the heart and pancreas, isolating a source of quickly dividing cells to heal a damaged organ has unmatched medical potential. However, finding these ephemeral cells is just the first hurdle; getting them to grow in adequate numbers is another story altogether. One way around this conundrum is to make them using embryonic stem cells. If a line of hESCs could be manipulated just so, an unlimited source of tissue-specific progenitors could be grown for medical purposes and used as a substitute for the more slowly dividing and exceedingly finicky adult stem cells.

Progenitor cells tend to confuse stem cell hunters because any sample of cells taken from the body contains a mishmash of stem cells, progenitor cells, and terminally differentiated cells. Telling them apart in a few cc's of blood or a gram of tissue is difficult because their differences are not strong enough to produce obvious physical cues.

WE KNOW WHERE YOU LIVE

Stem cells are found in a startling range of places. In some cases, adult stem cells remain elusive. The presence of their offspring, progenitor cells, and the fact that organs repair themselves are evidence they exist. On the next page is a drawing of interesting adult stem cell "breeds" and where they are found in the body. Adult stem cells reside beneath the skin; in the placenta and umbilical cord; and inside the olfactory bulbs, a pair of Q-tip-shaped organs that sit just behind the eyes. Bustling areas of stem cell activity are inside bones and behind the wall of the intestines. Even the connective tissue of muscles has stem cells within it. The organs, too: heart, lungs, liver, brain, and kidneys.

Most of our knowledge of adult stem cells comes from the hematopoietic stem cells, or HSCs. The most widely distributed stem cells, they make their primary home in the bone marrow. HSCs also hang out in the spleen, liver, lymph nodes, thymus, umbilical cord, and circulating blood. Titans among multipotent cells, the HSC forms nine kinds of blood cells. The epicenter of blood cell manufacture is in the

a human atlas of adult stem cells

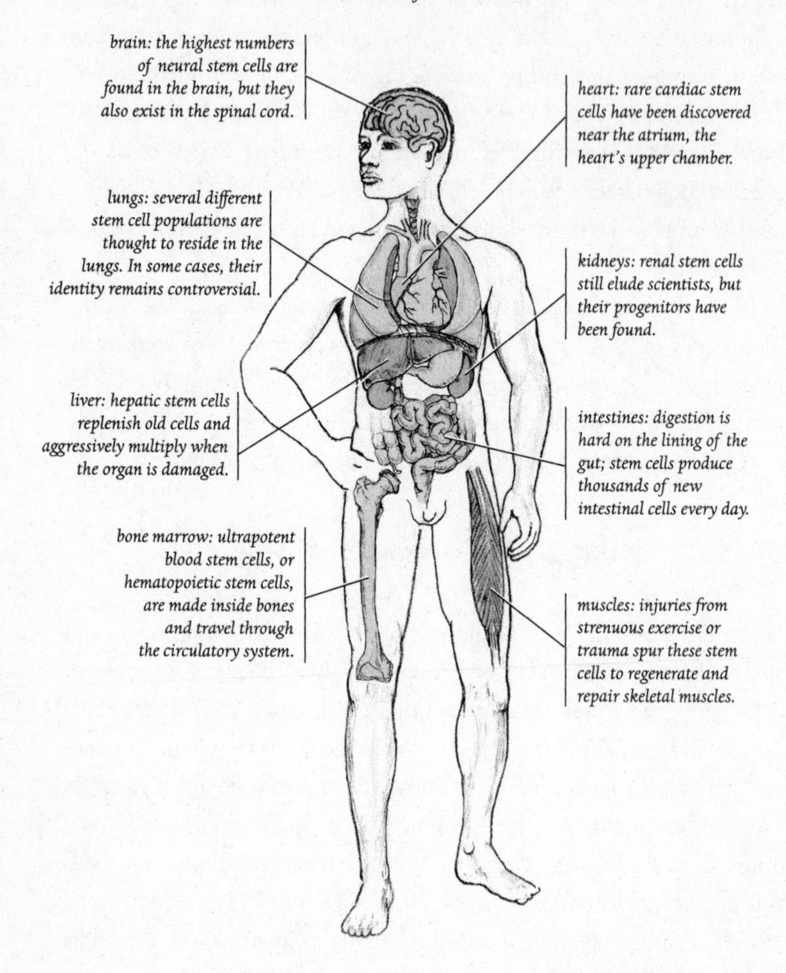

spongy spaces near the ends of the bones. There, swarms of HSCs and their progenitors produce platelets (the small cells that aid blood clotting), red blood cells, and a variety of immune or white blood cells. As shown in the illustration, it's a crowded, active place, with stem cells nestling next to the inner surfaces of the bone, where processions of progenitors and a multitude of newly-minted red blood cells prepare to exit the marrow into the bloodstream.

hematopoietic stem cells and their offspring

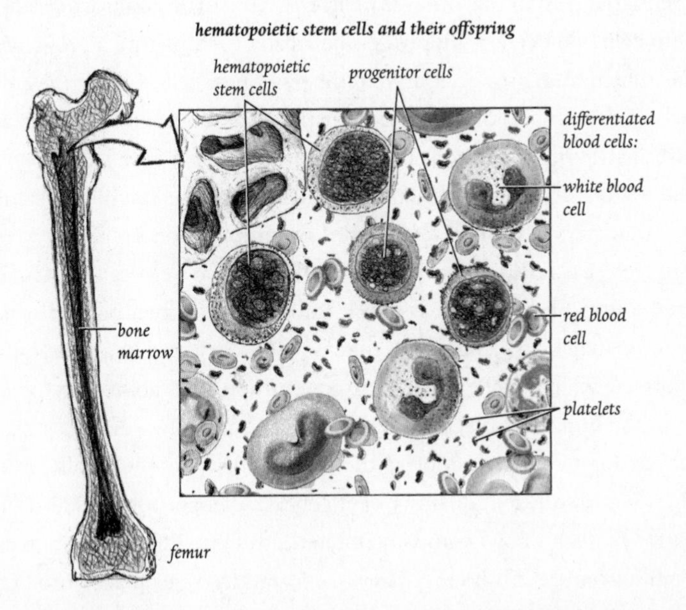

Red blood cells are short-lived, so their rate of replacement is prodigious—hematopoietic stem cells must make between 100 and 500 billion of them every day. When it comes to forming progenitor cells, blood stem cells again take the prize: one HSC will go through six stages of change before it finally differentiates into a red blood cell. As a consequence of vast amounts of research, blood stem cells were first to be used therapeutically, and new breakthroughs are introduced in Chapter 7. The mesenchymal stem cell is a multipotent, jack-of-all-trades stem cell also found in the bone marrow. Clinical research finds that adding fresh mesenchymal stem cells to blood stem cells during bone marrow transplants reduces the life-threatening immune reaction and increases their assimilation.

Besides the bone marrow, hotbeds of stem cell activity are located in the intestine and skin. We lose skin cells by the millions every day, so a rich source of progenitors must continuously renew the terminally differentiated cells of our outer skin layer, or epidermis. Like other adult stem cells, skin cells multiply geometrically, first making a progenitor cell, which then produces large numbers of new cells. Unlike the stem cells of the blood, however, the dermal and intestinal stem cells are part of an easily visualized and organized tissue structure.

A central assertion of twentieth-century mammalian neuroscience was that nerves cannot regenerate, the implication being no nerve stem cells exist. But such a stem cell called neural stem cell (NSC), has been found, and the hope is that it can be used to treat brain disease, nerve damage, and spinal cord injury. NSCs are found in a wide range of places, which may explain their uncommon adaptability—removed from the brain and placed in remote locations of the nervous system they settle in quickly, adopting the characteristics of the locals.[6] Finding NSCs in sufficient quantity has been a challenge, but in 2000, a California biotechnology company isolated a multipotent neural stem cell from human brain tissue.[7] They are found in other places too. The neurons of olfaction, or smell, are continually replaced by NSCs, and some forms of brain cancer are thought to arise from dysfunctional neural stem cells.[8]

Stem cells probably reside in most organs, and some are regenerative powerhouses. It turns out that mythology had a dose of stem cell reality: Prometheus looked down to find his eagle-ingested liver had miraculously healed. Damage to the liver, even removing half of it during surgery, mobilizes millions of stem cells to rebuild the organ in as little as a few weeks. Even relatively nonregenerative organs like the lung, heart, and kidney have some capacity to repair themselves, suggesting a nascent stem cell source. In fact, scientists have found progenitor cells in the lungs of mice and humans that change into smooth and shiny pulmonary muscle.[9] The cost of caring for diabetes and its increasing incidence means billions of healthcare dollars at stake—so the hunt for regenerative pancreatic stem cells is on. In less turbulent pockets of the heart, a slow, asymmetrically dividing population of stem cells has been discovered.[10] Adult stem cells can be found in the muscles, pancreas, blood vessels, and breast. Others show up in unorthodox places, such as midway through the hair follicle, deep inside teeth, and in fatty tissue. Just recently a source of blood stem cells has been isolated in human placentas.[11]

Because adult stem cells gravitate to their own microenvironments, their immediate family members—the progenitor cells—tend to as well. But it is unclear whether organ- and tissue-specific stem cells actually originate in the organs and tissues they are found in or move there as descendents of cells circulating in other parts of the body. There are other questions. Which stem and progenitor cells are best for medical uses? Which stem cells are the most flexible in the lab? Which are the most numerous, easiest to identify, and are most amenable to isolation? We will address these questions and others in the next chapter.

HABITAT: THE NICHE

The real-estate saying, location, location, location, also applies to adult stem cells, which are most productive in their home environment, the niche. If the lungs, skin, and liver are "cities" of cells, the niche is the neighborhood within, the specific environment where stem cells can be found. The space deep inside the bone marrow (shown in the illustration on page 67) is one example of a niche. And like any good neighborhood, the local environs provide services to ensure survival and protection. They also entice stem cells to stay put. Inside the niche, stem cells put their genes for division and specialization on hold.

How important is the niche? Things go badly without it. An embryonic stem cell removed from a mouse blastocyst and transplanted into an adult animal's abdomen unleashes a torrent of unregulated activity and becomes a tumor. Put the embryonic cell back in its natural habitat—as embryologists Beatrice Mintz and Martin Evans did in the 1970s—and peace and happiness reigns, and the cell develops normally. The niche marshals resources in an emergency. The local cellular environs of muscle fibers erupt into a hotbed of regeneration if nearby cells and tissues are damaged. Cell-to-cell contact is also important. In the niche of the fly gonads (a vanishingly small place), escort cells linked to stem cells shuttle them to and fro.[12] Stem cells lose their stemness if they are separated from the support cells that touch them. Once contact is lost with near neighbors, blood stem cells leave the bone marrow and enter the blood vessels. If conditions change in the niche or a stem cell detects signals from outside the niche, wanderlust sets in: stem cells begin to divide and roam.

Understanding the niche's inner workings shows promise for new therapies. For example, knowing which cells and signals are critical to wound repair could lead to drugs that speed healing after surgery or injury. Cancers like leukemia arise in populations of progenitor cells that reside in the bone's niche. When the cancer cells are removed from the environment of the niche, they die. Creating an artificial niche for stem cells to grow in could provide a controlled laboratory environment in which drugs that disrupt favorable conditions for cancer cells can be tested.

STEM CELLS IN ACTION

Cell turnover is especially high in the small intestine. Many thousands of new cells are needed daily in order for the gut to survive the local acidic conditions or the gustatory onslaught ingested by a lover of spicy food.

The small intestine is organized into involutions of tissue, called villi, and crypts, cavelike indentations at the bottom of each villus. Each villus has a ring of specialized stem cells near the bottom of each crypt. The crypt is the intestinal niche, and the diagram here shows the niche of the human gut in action.

the niche in action: the intestinal crypt

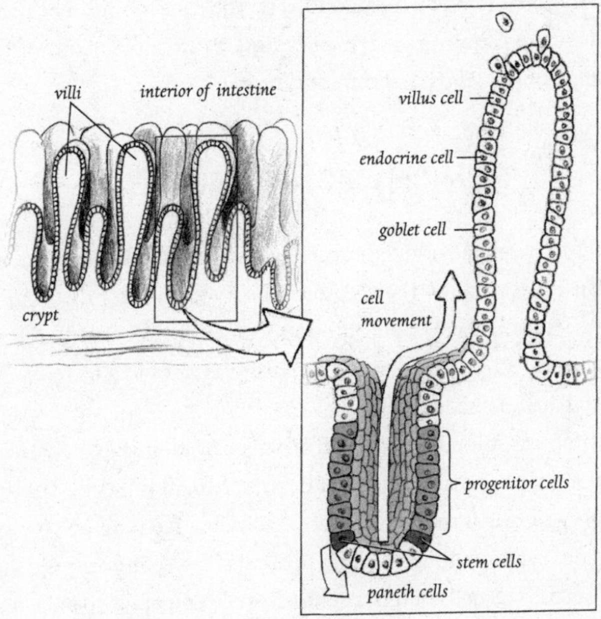

Stem cells that reside in the dark reaches of the crypt are multipotent and divide asymmetrically, forming different kinds of intestinal cells. Also present are support cells, which emit signals that regulate stem cell activity. To secure an adequate supply of mature cells at the top of the crypt, the stem cell first makes a population of rapidly dividing progenitor cells. The progenitors move up and out of the crypt, changing into mucus-making goblet cells, food-absorbing villus cells, and hormone-secreting endocrine cells. The stem cells also generate the nondividing paneth cells, which line the crypt's very bottom. After these cells reach their final resting spots, they eventually die from old age and are replaced by new cells migrating from above and below.

To replace dying intestinal cells, the crypt stem cells divide asymmetrically. Each cell division produces both a more differentiated cell type and the fully multipotent stem cell needed to keep the cycle going. The cycle from bottom to top is quite rapid: transit time from the bottom of the crypt to the apex of the villus is about four days.

STEM CELL SIGNATURES

The array of closely related adult stem cells presents a vexing problem for biologists. They are so alike that regular observation can't distinguish between them. Because genes play an important role in cell differentiation, cataloging cells by their genetic signature is one way to tell a cell from its immediate ancestor.

Such a profile can be constructed from special identifying markers. Cells have proteins called antigens that protrude from the cell's surface. Surface antigens perform many functions, including latching on to other cells, serving as receptors for incoming protein signals, and transporting amino acids across the cell membrane. Each cell type has a specific antigen profile. For example, all blood cells in the body have a marker called CD45 (CD stands for "cluster designation"). In contrast, all muscle stem cells express a different marker, CD56. The progenitor cells that make antibodies and white blood cells have markers CD10 and CD124.[13] But some surface antigens are found on both immature and mature cells, confounding the analysis. Because cells communicate dynamically, a high-strength cellular signal at one point will drop to low ebb a short time later. Also, it is possible that a cell expressing an antigen when it is in one location in the body (say, the niche) expresses a different marker when it is somewhere else. The trick is to find enough markers that spell a combination unique to that type of cell.

To stake out a claim on a new kind of stem cell, a scientist must first prove what it is not. To keep things straight, cells that *don't* bear the blood cell marker CD45 are called CD45 negative, or CD45(−), a nomenclature also used for other antigen genes. Presence of a gene, and evidence of its lineage, is represented by a positive sign (+). In the fuzzy purgatory of a stem cell hunt, researchers often speak of their discoveries only by surface antigen signature. "I found a Sca-1(+)Lin(−) CD45(−)CD31(−) cell," a stem cell hunter proudly says when announcing a kind of cell found in the blood. "That's nothing," another replies. "We've found a muscle stem cell with the signature CD45(−)Mac-1(−)Sca-1(−)CXCR4(+)beta1-integrin(+)!"

IS IT REALLY A STEM CELL?

How do we measure an embryonic stem cell for pluripotency, its particular brand of stemness? In mice and other animals, there are two standard tests. One such test transplants the cell back into an adult animal to see if it forms a teratoma. If a tumor with tissues from all the germ layers results, the cell is deemed pluripotent. Alternatively, a suspect ESC with a genetic marker is injected into the early embryo. If it's a bona fide embryonic stem cell, it participates fully in embryogenesis and the proteins made by the gene show up in different places in the animal. A test for a human embryonic stem cell uses a mouse without an immune system. Once it is transplanted into the animal, the cell will form an embryo-like structure with three germ layers.

Discerning adult stem cells from progenitor cells is not as straightforward. Testing for a unique signature, such as a cell surface protein, can help. If the protein is identified visually through specialized equipment, then the cells can be isolated and grown for use in human patients. Furthermore, a key signature of all stem cells is the presence of the enzyme telomerase, the same protein expressed by cancer cells (which causes them to grow out of control). In humans, only stem cells appear to have telomerase, supporting the notion that many tumors may start in stem cells as a first step in the progression to cancer.

When identifying adult stem cells, researchers look for three attributes that contribute to stemness: (1) stem cells renew themselves; (2) they differentiate; and (3) when transplanted, they settle in and function fully in their new environments.

Identifying and characterizing stem cells is a first step on the long road to their eventual therapeutic use. The next step involves observing and testing stem cells in controlled conditions designed to predict whether they can live up to their medical promise. The results of this research often make headlines. One thing is certain, stem cell science is young, and progress at this stage is rapid and full of debate—especially among scientists. One question is at the forefront of the debate: which has more potential to cure human disease, adult or embryonic stem cells? The next chapter addresses this question by sampling the major scientific breakthroughs at the edge of biology's newest frontier.

6

Seven Questions

Scientists Grow Adult Stem Cells from Nose![1]
REUTERS NEWS REPORT

Every year, hundreds of stem cell research papers appear in scientific journals.

Press headlines invariably follow. Close on the heels of a *Science* paper, a popular website announces, "Scientists Clone First Human Embryo."[2] An intriguing result published in *Nature Biotechnology* prompts an online news feed to declare, "New Hope for Hairless Seniors—Even Pick Your Color."[3] Newspaper and magazine articles tantalize with reports of new discoveries and what they mean for curing disease. The information comes fast and furious in part because stem cell research is moving at such a rapid clip.

In most cases, science journals report the results of experiments designed to answer the most pressing questions facing stem cell biology. A number of questions weigh on researcher's minds, some of which we have encountered in previous chapters. Can nuclear transfer make human cells for therapies? Why is the niche so important? How powerful is a newly discovered adult stem cell? Can stem cells help cure Parkinson's disease? Is there a stem cell that can mitigate the effects of diabetes?

What are the connections between stem cells and cancer? And perhaps the most intriguing question: how potent are adult stem cells? This prospect fuels articles like this one: "Scientists Grow Adult Stem Cells from Nose," which reports the discovery of an adult stem cell with embryonic stem cell qualities. If true, such a noncontroversial source of stem cells might repair any failing part of the body.

The questions that pique the public's interest focus on curing disabilities and disease, but the research casts a much wider net, focusing on small parts of a much bigger picture. A sampling of the current science illustrates how tantalizing pieces of the puzzle are falling into place. This chapter highlights the groundbreaking research of seven teams of stem cell scientists. The first three experiments feature embryonic stem cells and show how the cell lines can answer essential questions about disease. The final four investigate the elusive adult stem cells. Laboratory experiments are designed to answer a question or set of questions. The central question for each experiment is posed first. As we shall see, the answers are sometimes surprising and unexpected.

CAN SPERM CELLS REVERT TO EMBRYONIC STEM CELLS?

A few days after James Thomson announced his human embryonic stem cell (hESC) discovery, researchers at Johns Hopkins University published another dramatic result. This time, the cell line came from tissue removed from five- to nine-week-old human fetuses donated after elective abortions. The lead author of the paper, John Gearhart, isolated the cells from an area called the gonadal ridge.[4] These primordial cells eventually become oocytes and spermatocytes, the predecessors of eggs and sperm.

Gearhart noticed his cells looked and acted very much like hESCs, so he called them embryonic germ cells. They formed embryoid bodies, the curious constructions observed by Leroy Stevens. What's more, they sported the gene signature of embryonic stem cells and were quite potent, changing into three different germ layers. One major difference was staying power. Human embryonic stem cells could make countless copies of themselves, but embryonic germ cells could not.

Using fetal tissue, human eggs, and embryos for research is controversial, so scientists are exploring ways that would use them more sparingly. Because embryonic germ cells had stem cell powers, it made sense to start with one of the cell types further along this pathway. If a stem cell in the germline of an adult animal was pluripotent, the method could make lines of embryonic stem cells without using embryos.

In Göttingen, Germany, a group led by Gerd Hasenfuss took on the challenge.[5] Not surprisingly, they picked an adult male mouse, and went after its testicles. They found a stem cell there responsible for making sperm throughout life. After perfecting the culture conditions, the group made long-lived lines that morph into different cell types, including beating heart cells. When they marked a cell from one the lines and put it into an early mouse embryo, they found its progeny in various organs, including sperm and eggs. The cells had normal numbers of chromosomes and expressed the right gene signature. The impressive results were published in the April 2006 edition of *Nature*—Hasenfuss had discovered pluripotent cells that had passed all the embryonic stem cell tests with flying colors.

Do the Germans have a new, noncontroversial source of embryonic stem cells? If there is an equivalent cell in man, then converting them into immortal lines could mean a supply of replacement cells for a variety of ailments. Unlike other adult stem cells, the starting material is plentiful and could be a possible alternative to somatic cell nuclear transfer (SCNT), which would need donated human eggs in order to make custom-made cells and tissue for patients. The German discovery was watched closely, coming on the heels of one of the biggest scientific scandals in history. A lab in South Korea reported they had accomplished nuclear transfer using human cells, but the experiments were later found to be fake (see Chapter 10).

While the discovery is exciting, more work must be done. Stem cell hunters know that many features of mice go missing in men. The cells must be followed to see if they remain normal over long periods of time. Sperm cells bear genetic imprints that may not be easily erased in the laboratory. Finally, a female source of cells is needed. Girls are born with all the egg cells they will ever have, which suggests that the female equivalents of these cells don't exist. But a Harvard team last year found evidence of egg production in adult mice, suggesting that egg precursor cells may be housed in women's ovaries.

CAN EMBRYONIC STEM CELLS CURE PARKINSON'S DISEASE?

Time will tell whether experiments show that lines of embryonic stem cells can be made using nuclear transfer. The next challenge will try to produce a specific cell type from a donated IVF embryo. Parkinson's disease is a brain disorder caused by the death or impairment of neurons. Normally, these cells produce a vital chemical known as dopamine. Dopamine ensures smooth, coordinated movement of the body's muscles. When approximately 80 percent of the dopamine-producing cells are damaged, tremors, slowness of movement, and muscle rigidity results. Though Parkinson's disease can be treated, it cannot be cured. Cell transplants from fetal tissues can change the course of the disease, but these cells don't reliably make the required numbers of dopamine-synthesizing neurons. In contrast, embryonic stem cells (ESCs) not only can proliferate prodigiously, they can generate dopamine neurons. If ESCs are to become the basis for cell therapies,

scientists must develop
can demonstrate that
must work out the me

Research to treat P
tory rats. Biologists a
the brains of rats
destroyed.[6] Of 25
producing dopami
drawback of using
is that once remo
benign tumors;
brain disease. In

A few months later,
the National Institutes of Health (NIH)
same rat model, but this time they attempted to fine-tune the
tion of ESCs prior to transplanting them.[7] They reasoned that nudging ESCs to become cells that produce dopamine *before* transplanting them would avoid the problem of tumor formation, an important result to show before human clinical trials could proceed. The NIH team grew a concentrated population of midbrain neural stem cells from rat ESCs. From previous work they found they could manufacture the gamut of cells in a neural pathway: neural stem cells, neural progenitor cells, and mature neurons that produce two brain-signal transmitters, serotonin and dopamine. McKay and Kim also introduced special growth and cell differentiation agents to the cell culture medium that increased the numbers and desired types of cells. The engineered cells were anatomically, neurochemically, and physiologically identical to normal neurons. Once implanted, they secreted dopamine, extended long axons to reach other neural cells, and formed synaptic connections. Like the rats in the Harvard study, these rats showed a marked behavioral improvement, but without the tumor troubles.

Tweaking lab culture conditions to make just the right kind and kinson's disease and other maladies like are to be helped with stem cells. In 2005, University Medical Center in Jerusalem took g hESCs into rats that mimicked Parkinson's dis- grafts survived for three months and produced the r dopamine. A gradual improvement in rat's behavior d no tumors developed.[8]

CAN CANCER BE REVERSED?

Nuclear transfer is an amazing biological trick, and Rudolf Jaenisch has trouble believing what he is seeing. Standing eye-to-eye with a cloned animal does little to persuade him. For a man who has studied biology for more than 40 years, his voice registers awe when he says, "I still cannot believe this. I am amazed it works as well as it does."[9] He refers to an empty egg, a gelatinous pinprick with no nucleus. Give it a nucleus and it will go about making an animal that looks you in the eye.

Jaenisch has an easygoing way and talks in a question-and-answer style that reveals his passionate curiosity about biology. "What happens in nuclear transfer?" he begins. "You take a nucleus that has been programmed in a way that is right for its function. In Dolly's case, it came from a mammary gland—there to produce milk. The genes essential for embryonic development are present but silent. Then you transfer the nucleus and ask the egg to reverse in a matter of hours what happens over months or decades during normal development of a cell. The silent genes speak and cleavage begins and all the rest. It is amazing, I tell you!"

A founder of the Whitehead Institute at MIT, Jaenisch was the first to develop a transgenic mouse. He has produced many mouse models of human disease and remarks off-handedly, "You know, cancer is a side interest of mine." His wonderment at the egg's reprogramming prowess has him scratching his head about a related phenomenon in cancer. Cancer is caused by faulty genetics: key genes in healthy cells mutate and, as mutations multiply, tumors grow. But there is an epigenetic (or "outside" genetic) component of cancer that silences genes without altering the sequence of DNA. If an otherwise active gene that keeps cancer in check is silenced, tumors result. Some cancers have combinations of genetic and epigenetic effects. For example, gene silencing alters 85 percent of the genes responsible for colon cancer. In breast cancer, nearly half the genes are quiet. Jaenisch believes if an egg can reverse—or reprogram—silent developmental genes then perhaps the epigenetic effects of cancer can be reversed, too.

"How much of cancer is reversible?" he asks. This question reveals another—how much of cancer is due to genetic causes and how much is due to epigenetic causes? To answer this, Jaenisch and his collaborators Konrad Hochedlinger and Robert Blelloch used SCNT to transfer nuclei of different kinds of cancers into the enucleated eggs of lab mice. Most of the developing embryos died or became tumors. But cloned blastocysts made from melanoma cells appeared to be normal. The stem cells were removed and put into a normal mouse embryo made by fertilization. The cells incorporated into the germ layers and went on to make most of the tissues of an adult mouse. If the cloned blastocyst was transplanted back into a mother mouse, it developed into a normal baby mouse. Although the egg reprogrammed the silenced genes, it could not reprogram the genetic cause of melanoma: eventually these mice developed cancer.[10] "This settles a basic biological question," says Jaenisch. "The epigenetic elements of cancer are reversible; the genetic elements are not." Jaenisch expects that drugs that reverse the epigenetic components of cancer could treat and slow the disease.

Like a cancer cell, a stem cell is immortal, living as long as the creature does. Jaenisch believes there is a stem cell link to cancer, and adult stem cells may be particularly susceptible to cancer's deadly grasp.

"Cancer involves many changes to DNA, and these changes accumulate over time. Now, alterations add up in an immortal cell but not so much in a mortal cell—there is not enough time to accumulate the effects," he says. "When a stem cell in its niche is finally switched to a cancer cell, the requirements of the niche may be disrupted. The cancer cell may escape, produce cancerous progenitor cells, and before you know it, you have a tumor."

The German-born scientist is just as excited about using SCNT to cure other diseases, such as Parkinson's and diabetes. Can cell transplantation cure these diseases? Without hesitation, he replies, "The answer is a solid yes." To back up the claim, his MIT lab used a combination of SCNT and gene therapy to cure a mouse suffering from severe combined immunodeficiency (SCID), known in humans as the "bubble-boy disease."[11] Like their human analog, SCID mice are missing the ability to make white blood cells. The lab made a blastocyst using a nucleus from a skin cell of a SCID mouse. The ESCs inside still had the defective SCID gene, so they repaired it with a working version from a healthy mouse. They then made an ESC line—adding the right cocktail of chemicals caused the embryonic cells to differentiate into blood stem cells. The final step was a mouse version of stem cell therapy: the cells were injected, colonized the bone marrow, made healthy white blood cells, and rescued the mouse.

Using nuclear transfer for therapeutic purposes is intriguing, but it is not yet practical to engineer hESCs to transplant into every sick patient. In the near future, the lines have a much more powerful purpose. Jaenisch describes the potential to mimic human disease in a laboratory setting as "enormous." An hESC line made from a nucleus taken from the pancreas of a type I diabetic unable to produce insulin can be compared to a normal line of insulin-producing cells. How many genes are involved? Which proteins are responsible? Can drugs reverse the effects? The questions Rudolph Jaenisch asks foretell the discoveries he's pursuing.

CAN STEM CELLS CURE DIABETES?

Many adult stem cells have not yet been positively identified. Whether they can be found—or exist at all—are important questions when major organs are involved. One such organ is the pancreas. After a big meal, the body stores excess sugars as energy reserves. A specialized cell in the pancreas called the β-cell (beta cell) senses the abundance of sugars and secretes the hormone insulin. In turn, insulin causes cells throughout the body to take up the abundance of glucose and store it for later use. Worldwide, over 5 million young people have diabetes (type I or juvenile diabetes). They are plagued by a faulty immune system that destroys their own β-cells, requiring a lifelong dependency on injected insulin.[12] The good news is, cell therapy can cure type I diabetes. Transplanting clusters of pancreatic cells (called islets) containing β-cells (obtained from donated organs) into patients permanently eliminates the need for insulin injections. But the cure comes at a price. Pancreatic cells from several cadavers are often needed for each transplant, and a regimen of powerful immune suppressants is required to keep the cells from being rejected. Does the pancreas have adult stem cells that replenish insulin-producing β-cells? If so, they could be harnessed to use as therapies for diabetics.

Yuval Dor and Douglas Melton, scientists at the Howard Hughes Medical Institute at Harvard University, decided they would look for evidence of pancreatic stem cells in the laboratory mouse.[13] Their experiment relied on an assumption about how stem cells repair and maintain organs. For example, epidermal stem cells slowly replenish cells of the skin as they die and divide rapidly to mend a flesh wound. Dor and Melton imagined that pancreatic stem cells behave much in the same way, that they occasionally replace elderly β-cells and become livelier after an injury. To measure how quickly pancreatic cells are replaced, they indelibly marked the DNA of insulin-producing β-cells, as diagrammed on the next page. Once marked, the researchers could track the β-cells as they grew old and died. Adding a

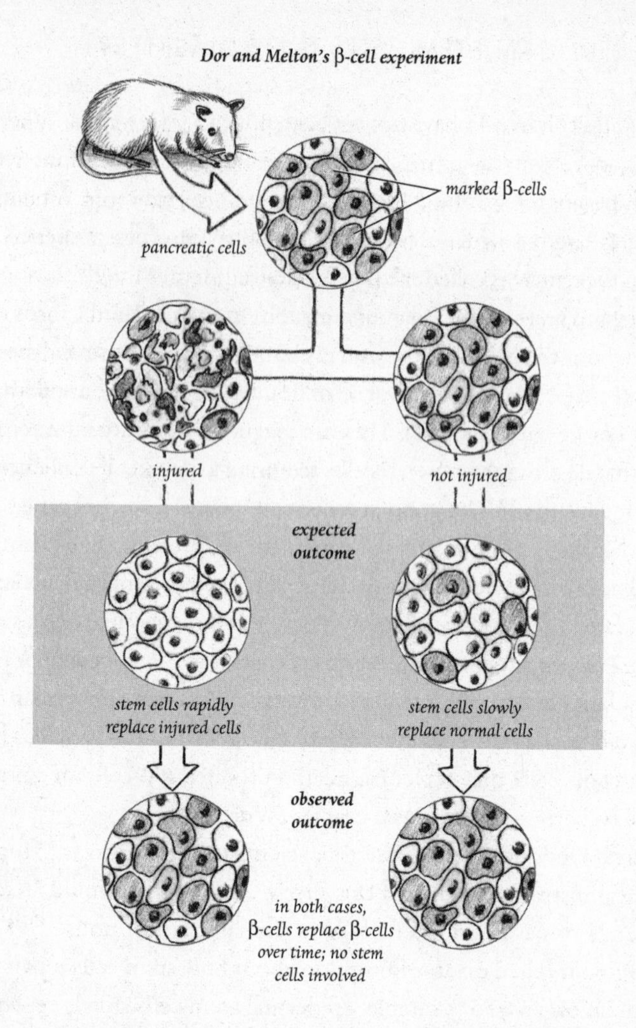

hormone caused the engineered β-cells to change color. The team wagered that over time the old, marked cells would be replaced by new, unmarked cells coming from stem cells. In a normal pancreas, the rate of replacement would be slow. In an injured pancreas, the marked cells would disappear much more rapidly. Surprisingly, neither result occurred.

What the researchers observed after simulating an injury was no change in the number of marked cells, even up to one year later. As the cells replenished the islets, new cells with the same genetic marker appeared in their place. It turns out that new cells don't come from stem cells at all; they come by way of normal cell division. When an old β-cell dies, it is replaced by a remaining β-cell that divides in two.

Dor and Melton's result suggests that adult stem cells may not repair the pancreas and that therapies for diabetes should focus on ways to make more β-cells. Could β-cells be removed from cadavers, multiplied in the laboratory, and injected into patients? Yes, but putting cells from a middle-aged cadaver into a five-year-old child could complicate the child's health later in life. The results of the Harvard experiment make a case for pursuing a source of β-cells made by way of embryonic stem cells. It also illustrates how adult and embryonic stem cell research goes hand-in-hand. An hESC line made using the nucleus taken from a child's somatic cell could make a lifelong culture of genetically matched insulin-producing cells.[14]

ARE BLOOD STEM CELLS MULTIPOTENT?

A long-held dictum of adult stem cell biology says that once a cell commits to the lineage of its germ layer, it cannot change. A second principle, just as time-worn, asserts that adult stem cells are limited in what they can become: once a blood stem cell, always a blood stem cell. Yet a third states that stem cells found in a specific tissue are there only to replenish the cells of that tissue. Six years ago, a flurry of research suggested that all three were suspect: blood stem cells could be injected into other places in the body where they appeared to make new heart, lung, and liver cells and muscles, nerves, bone, and capillaries.

Yuehua Jiang and Catherine Verfaillie at the Stem Cell Institute of the University of Minnesota wondered why so many laboratories reported this startling plasticity.[15] Perhaps they weren't peering at blood stem cells at all. Could there instead be extremely rare adult stem cells in the bone marrow with powers that rival those of embryonic cells?

The Minnesota group derived a rare variety of stem cell found in the bone marrow, brains, and muscles of humans, dogs, rats, and mice. They named them multipotent adult progenitor cells (or MAPCs). In the lab, MAPCs are quite powerful. They grow indefinitely and express telomerase, the enzyme associated with immortality. They appear to change into the cells of all three germ layers and go on to make liver cells, neurons, and endothelium, the cells that line blood vessels and body cavities. MAPCs and embryonic stem cells have very similar gene signatures, but MAPC genes express the proteins at much lower levels. A MAPC passes a crucial embryonic stem cell test: put one inside a mouse embryo and it integrates seamlessly with embryonic stem cells, eventually contributing to all the parts of a baby mouse. These cells are exceptionally rare: current estimates indicate that less than 2,000 MAPCs exist in a single adult mouse. They are finicky to grow, too. The criticism of the Jiang/Verfaillie experiments is that other labs have found it difficult to reproduce their results.

That said, the Jiang/Verfaillie discovery may be evidence of the existence of an adult stem cell with embryonic stem cell potential. MAPCs may in fact be "universal" adult stem cells. Because they are observed only in the laboratory, the jury is still out on whether these cells actually exist in the body and are as powerful as they seem. If they are everything they could be, they represent an exciting discovery.

HOW IMPORTANT IS THE NICHE?

Searching the hideouts of secretive adult stem cells may tell scientists what makes them multiply. As a result, people who suffer from baldness may soon be pinning their hirsute hopes on the hair follicle, found just below the skin. Hair regenerates—anyone who plucks a stray hair from an eyebrow knows it stubbornly returns a few days later. Suspicions are that hair renewal (and permanent hair loss) depends on stem cell activity. In 2004, researchers at the University of Pennsylvania isolated adult stem cells from inside the follicles of mice.[16] They transplanted the cells into hairless spots on test mice, and four weeks later, hair began to grow. Another lab developed a line of follicle cells made from embryonic stem cells. The cell culture experiments were designed to uncover the developmental steps (and genes) in hair formation, including the creation of progenitor cells, the prolific dynamos that make mature hair follicles.[17]

If hunting stem cells is the aim, then look for the niche. That's what niche specialists Tudorita Tumbar, Elaine Fuchs, and colleagues at Rockefeller University did when they designed an experiment to answer three questions.[18] What are the features of the follicle's niche? Which cells in the niche are important for replacing hair? What signals do stem cells use to regenerate hair?

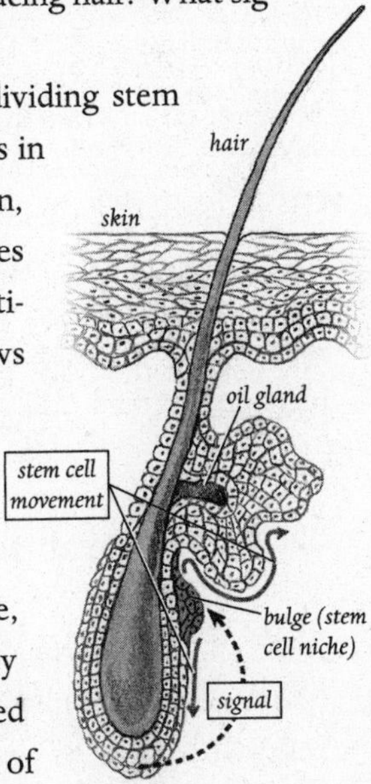

The Rockefeller team knew that slowly dividing stem cells tend to stay in the niche. But which cells in the niche are stem cells? To answer this question, they engineered mice with a gene that expresses a light-activated marker, a protein easily identified under a microscope. The protein glows most brightly when cells are barely dividing. In contrast, rapidly dividing cells dilute the marker until the cells barely glow. When they looked at the follicle under a microscope, all the brightly lit cells lived in the bulge, a staging area half way up the hair shaft. They found mRNA in the cells (the molecule involved in making proteins) that matched mRNAs of other kinds of stem cells. The group also identified a constellation of genes that appear to turn on and off during hair regeneration.

Tracking stem cells shows how hair renews. The illustration here shows the anatomy of a hair follicle. The stem cell niche is in the bulge, a swelling of cells midway up the follicle. The stem cells in the niche begin to divide when signaled to do so by cells at the base of the follicle in the hair bulb. When a hair is lost, the follicle temporarily recedes, causing the cells in the hair bulb to send signals. After receiving the message, only a few stem cells spring into action and begin to divide. Duplicate daughter cells stay behind in the niche. The remaining cells exit and travel downward and upward, dividing and differentiating to rebuild the follicle, which in turn grows a new shaft of hair.

When stem cells become permanently quiet, baldness results. Urging these recalcitrant cells out of their catatonia may mean shiny pates will grow hair again. (Despite the news report mentioned in this chapter's introduction, it will be even longer before hairless seniors can pick their favorite color.) The research also suggests medical uses. Skin grafts lose their ability to grow hair, so stem cell therapies may someday help burn victims. Most importantly, skin stem cells aren't restricted to hair renewal. Fuchs and Tumbar found when skin is injured, stem cells leave the niche in droves to repair the wound.[19] Defining the niche, describing a cell's "stemness," and listing the gene signatures are essential steps before medical uses for wound repair, regenerative medicine, and skin cancer can be developed.

ARE ADULT STEM CELLS PLASTIC?

Unlike the boisterous declamations of members of Parliament or the chaotic chair throwing of the Japanese Diet, when scientists have a bone to pick, they are mostly polite. The questions and comments at scientific meetings are usually prefaced with "with all due respect" or "thank you for the interesting talk," but what inevitably follows is the cold calculus of someone who knows how to zero in on a weak spot. "How do you explain your result in light of evidence that shows otherwise?" they wonder. Or, "Couldn't your data be explained another way?" A big issue at these meetings is whether cells ignore the conventional wisdom of developmental biology. Cell differentiation, says biological dogma, is overwhelmingly in one direction and is progressively restrictive. A stem fated to make neurons cannot make blood cells; a stem cell fated to make white and red blood cells cannot make a heart or a liver.

In the late 1990s, a series of research papers challenged how scientists think about cell differentiation. Mixed populations of cells taken from the bone marrow containing hematopoietic stem cells (HSCs) were first tagged with a molecular marker and then injected into mice that had been irradiated or had injured organs and tissues. Like microscopic rescue workers, the adult stem cells appeared to go to the site of the injury and ignored their germ line heritage. They seemed to change into a mélange of tissues: skin, lung, intestine, kidney, liver, pancreas, muscle, blood vessels, heart, and brain. Most astonishing were HSCs that could switch cell lineages and become heart cells—offering a potential cure for heart disease. If these results weren't startling enough, several laboratories claimed that other adult stem cells could perform the same feat. Neural stem cells, they said, could do an about-face and make blood cells and immune cells.

Though biologists are unanimous that even the most potent adult stem cells cannot approach the therapeutic power of embryonic stem cells, the debate about whether cells transdifferentiate or dedifferentiate has been propelled into a political orbit. A new journal called *Stem Cell Reviews* dedicated its first issue to the phenomenon of plasticity. The journal's editor-in-chief Stewart Sell notes how politics and plasticity have become strange bedfellows. "The opponents of embryonic stem cell research argue that anything that can be done with embryonic stem cells can be accomplished with adult stem cells," he writes. "Proponents of embryonic stem cell research have used the negative data (that stem cells are not plastic) to argue that adult stem cells can never do what can be done with ESCs."[20] For most scientists debating the issue, the point is not whether one kind of cell is best to cure human disease—it is too early to tell. For them, the question is simple: Are adult stem cells plastic or not? When pressed for an answer, Amy Wagers, a Harvard University blood stem cell expert says, "We're still debating this question. It's too early to tell which way things will fall."[21]

Labs are reporting that adult stem cells continue to ignore the dogma of developmental biology. Researchers at Sweden's Karolinska Institute have reported blood cells switching to lung cells.[22] Some posit that progenitors of white blood cells can repair severely injured muscles. A team at Robert Wood Johnson Medical School in New Jersey led by neuroscientist Ira Black has isolated a subcategory of HSCs called stromal cells, and watched them change into neurons before injecting them back into fetal rats.[23] His team followed the fate of the cells for two months after the rat was born. When they examined the brain tissue under a florescence microscope, they saw thousands of neurons born from the blood cells. Proof, Black says, that adult stem cells from the blood travel, transdifferentiate, find new niches, and survive. At Yale University, Diane Krause and her colleagues have found evidence that cells found in the bone marrow differentiate into liver, lung, and cells covering internal surfaces of the body.[24] Krause believes she has a powerful yet uncharacterized "marrow-derived cell" on the order of Catherine Verfaillie's MAPC. "There is no question that a marrow-derived cell can make epithelial cells," she says.[25] Researchers at the University of Iowa report that a subset of stem cells found under the skin of mice change features in lab dishes and hang out in the bone marrow eight months later.[26] Other investigators have found that neural stem cells isolated from the brain do unexpected things in a culture dish. Add just the right ingredients and these cells produce insulin, a hormone associated with pancreatic cells. Alan Trounson, director of the Immunology and Stem Cell Laboratories at Australia's

Monash University, sums up his take on the debate in a recent survey of one international stem cell meeting. "The contribution of hematopoietic stem cells to tissues appears now to be irrefutable, although the underlying mechanism remains unclear."[27]

The number of research papers reporting plasticity outnumbers papers that dispute it by 10 to 1.[28] Stanford's Irving Weissman and Harvard's Amy Wagers—experts on blood-forming stem cells—are among the doubters. They designed an experiment in which they marked every cell in a mouse with a gene making a protein that glowed green.[29] Then, they took only one green stem cell and injected it into an unmarked mouse that had been treated with irradiation—a procedure that kills all the native bone marrow cells, including any stem cells. The green blood stem cells multiplied and spread throughout the mouse. If plasticity is at work, then the embryonic stem cells should change into green lung, skeletal, gut, brain, liver, and kidney cells. The pair looked at over 15 million cells and observed only eight green cells that had somehow lodged in the brain and liver. They suggested the eight cells were not due to stem cells changing fates but to rare instances of fusion: stem cells combining with resident brain and liver cells. Mostly the lone green HSC made blood—what blood-forming stem cells are supposed to make. So the duo then asked, "Why isn't plasticity part of the normal biological operating system of an animal? Like the egg's ability to reprogram DNA, plasticity may not be a biological principle but rather an experimental peculiarity. Wagers and Weissman combed through the literature and concluded that stem cell plasticity may be due to a number of other reasons:[30]

1. Many cells look alike. To see whether a stem cell has switched roles, you must first engineer it with a gene whose protein marker is easily detected and then follow it. Even then, proof is elusive. A powerful microscope must be used to tell whether you are looking at one cell or a cell that has somehow fused with another.

2. Few studies determined whether the transplanted cells actually repaired the injured tissue. The ultimate test of a stem cell's moxie is not what it looks like or where it lands but whether it delivers the goods—that is, whether it functions.
3. In many cases, a mixed population of blood cells was transplanted. Which cell made the switch? A blood stem cell switching to a neuron is interesting; a neural progenitor cell that turns into a neuron is not. Using a pure population of donor stem cells would help settle this question.
4. Though they are made in the bone marrow, HSCs end up in organs like the brain and muscles as they travel through the bloodstream. A group of muscle cells that changed into blood cells may have had blood stem cells among them in the first place.
5. The percentage of cells that switch fate is very low. The culprit could be an ultra-rare multipotent stem cell (such as the MAPC discovered by Verfaillie and Jiang), and not an HSC.
6. An experiment repeated in another's hands is confirmation of its claims. Other labs could not duplicate some early results.
7. In late 2002 and 2003, independent laboratories reported that neural stem cells fuse with bone marrow cells, producing cells with double the chromosome number. Other experiments found that though a damaged liver regenerates after an injection of HSCs, the effect is due to cell-to-cell fusion, not plasticity.[31] At Stanford University, Helen Blau reported that HSCs fuse into mouse muscle fibers and that the genetic signature of blood stem cells can be observed in the cerebellums of mice one year after transplantation.[32] The stem cell hadn't changed; it had just combined its genetic markers through fusion, confounding the

analysis. A year later, other results showed fusion of HSCs to neurons and heart cells. Fusion occurs naturally, too—the best example is fusion of the egg and sperm during fertilization.

The Wagers/Weissman evidence suggests that reports of adult stem cell plasticity may be due to other reasons than cells switching their fates. However, the fact that hematopoietic stem cells can help repair damaged organs is an important result and could lead to using stem cells as therapeutic agents. In any event, the discussions these experiments provoke are evidence of a vigorous and healthy research environment; lively debate is a key part of the fact-finding process.

FROM LAB BENCH TO BEDSIDE

The particular science of human stem cells is not yet a decade old. In the past, scientists observed the effects of stem cells without knowing all that much about their specific roles in human development and health. Observing how an embryo changes and develops into a fetus is one thing; knowing that stem cell differentiation is behind the change is quite another. The same goes for biomedical therapies. That some therapies are successful is apparent, and—this is key—some of the reasons behind the successes are *now* attributed to stem cells. Knowing more about how these cells work will surely improve existing therapies. In other instances, stem cells will themselves become therapies. In both cases, the good news here is that some patients are receiving certain treatments because scientists are asking questions like the ones we just explored. In the next chapter, we discuss how discoveries like these make their way into the clinic for human use.

7

The Future of Medicine

I want things to happen quickly. I certainly want to benefit within my lifetime. I don't want to get out of this wheelchair at the age of 75. I am 51, and am now very healthy, and would like to be out of the chair very soon. I'm not willing to resign myself to being an advocate for research that will benefit people only after I'm gone. I'm not that noble.[1]

CHRISTOPHER REEVE (1952–2004), IN A 2003 INTERVIEW

In the face of hard statistics, one wonders how modern medicine can help so much suffering. By 2010, over 2 million Americans are projected to contract end-stage renal disease, at an aggregate cost of $1 trillion. In 2001, nearly 80,000 people needed organ transplants, fewer than 24,000 got them, and 6,000 died waiting. Of those receiving organs, 40 percent die within the first three years after surgery.[2] One in five of our elders 65 years old or older will require temporary or permanent organ repair or replacement during their remaining years.[3] In 2002, the prevalence of diabetes in the United States exceeded 18 million people—6.3 percent of the population. That year, total heathcare costs of diabetes surpassed

$130 billion.[4] Cancer kills one out of four of us, more than 1,500 people a day.[5] Even though we are living longer, many octogenarians are unable to appreciate their lengthy lives: nearly half of the people over age 85 have Alzheimer's disease. American lifestyles promote physical inactivity and overeating, causing morbid obesity, hypertension, and diabetes. Add to this list crippling conditions such as spinal injury, Parkinson's disease, multiple sclerosis, AIDS, and a host of genetic and metabolic disorders.

Heart disease is the biggest health crisis of all. In 2004, more than a million Americans died from cardiac failure and stroke, and heart disease leads deaths by all causes, outpacing cancer by 40 percent. No longer does it afflict only the old. More than 64 million Americans suffer from it, but only 25 million are 65 years or older. The total cost of treating cardiovascular diseases and stroke in the United States in 2004 was estimated to reach $368 billion.[6]

Given an ever-widening chasm between treatment and morbidity, it is no wonder the stem cell has become a common denominator of hope. Behind the sobering facts, patients and their families ask, "Will there be a cure? And will it be in time for us?"

WHEN?

Much of the promise of stem cells rests on a scheme for replacing parts worn out by age, injury, or infirmity. Unfortunately, the reality of stem cell biology is overshadowed by the hype. For example, the future is imagined to hold an inexhaustible source of stem cells with a perfect genetic match banked at a local hospital, available for your every medical whim. Need a new pancreas? Place your order and, three weeks later, a new one lies ready and waiting in the surgical suite. Heart failure? No worries—a few injections with multipotent stem cells will grow new cardiac tissue. Thus, many twenty-first century patients are imagined to extend their lives—through a kind of patchwork medicine, held together by a fabulous, potent cell. This future sounds incredibly exciting. But it will take time—and vision—to us get there.

The truth of the matter is, we've got a good distance to go before regenerative medicine—a catchall term for stem cell therapy—will help large numbers of patients. It is very possible that many diseases will have to wait for cures from other quarters of medicine. Before any medical treatment (including cell and tissue transplants) is made available through hospitals or clinics, it must first be tested in humans through tightly regulated phases of clinical trials. The first phase determines safety and side effects in a few dozen subjects; the second phase tests efficacy in hundreds of patients; the third and subsequent phases try to prove statistical significance and confirm its effects in many hundreds or thousands of patients. The U.S. Food and Drug Administration (FDA) evaluates the data, and if the results pass muster, the product is approved for sale and moves to the market. Developing a new therapy goes slowly and is *terribly* expensive—discovering, testing, and manufacturing one new drug can take between 10 and 15 years and cost nearly a billion dollars.

A hypothetical timeline of a new treatment for skin transplants might look like this:

- *Basic Research*: In 2006, a source of powerful adult stem cells is discovered beneath human skin. The rare cells are fingerprinted by genetic markers, and the markers are used to isolate the cells from the body and culture them in the lab. Over the next two years, technology is developed to grow the cells in quantity and used to change them into a variety of skin cell types.

- *Preclinical Research*: Different lines of skin stem cells and their progenitors are transplanted into the injured skin of a transgenic mouse with no immune system (to prevent rejection of the human cells). Over time the transplants are observed. One line works: the cells survive, go to the site of the injury, integrate into the skin, and heal the wound. Other kinds of animals are similarly tested. The tests take three years to complete.

- Clinical Research: The encouraging results in animals prompt tests in humans. In patients with severe burns, the patient's own skin stem cells are cultured, multiplied, and then transplanted at the wound site. The cells improve blood flow, promote healing, and reduce scarring. Using adult stem cells is not the only way to approach the problem. An hESC line using nuclear transfer might also produce the skin stem cell in question. The technologies are further developed by companies, tested in more humans, and manufactured for use for burn victims. In 2014, the FDA approves the first cell therapy for use in clinics.

If the treatment being studied is for a disease with a genetic cause, another wrinkle must be ironed out. The faulty gene has to be corrected before the cells are reintroduced or the transplant could succumb with time, as did the original cells. This presents an added set of challenges to stem cell transplants. Once a genetically engineered stem cell is placed into the body and grafts into an organ, it may be there for life. If the change is in one of the wide-ranging cells of the blood or nervous system, the proteins made by the new gene will be everywhere in the body. Care must be taken to limit the effects of the therapy only to the affected areas.

HOW MUCH?

Customized treatments that can't rely on economies of scale are likely to be very expensive. For an adult stem cell regimen, the tissue in which the stem cells reside must be biopsied—perhaps more than once—surgeries that can put elderly patients at risk. For any cell therapy, the methods for isolating, growing, and expanding the cultures must be perfected—complications not yet perfected for adult stem cells. The procedures must produce millions upon millions of homogenous, long-lived cells that exhibit stemness. Like any transplant, the cells must be free of contamination with unwanted viral, bacterial, or chemical agents.

To avoid "homegrown" protocols and ensure quality, companies and hospitals will need to standardize laboratory, manufacturing, and clinical practices. Health professionals will need training to provide proper informed consent and oversight of the procedures. Some researchers assert that for each patient, between 10 and 20 technicians will need to work full-time in specialized laboratories. The costs for such individualized treatments, they say, would be astronomical.[7]

A different strategy may reduce the cost. Rather than developing a custom stem cell line for each person, nationwide banks of several thousand hESC lines could be developed.[8] The banks would use a test called HLA (histocompatibility antigens) typing to match donor and recipient genes, minimizing tissue rejection. The closer the HLA match (either from family members or from outside donors), the less the chance that rejection will be a problem. A similar list of donors already exists. Over 6.5 million individuals have already been HLA-typed for bone marrow registries.[9]

Other experts contend that individual treatments are feasible, and that once competition heats up, market forces will conspire to bring down prices. If a stem cell therapy can cure, they argue, then all the downstream costs of caring for chronic illness go away. A high initial price for injecting stem cells would be more than offset by future medical savings.

However, even with the concerns of time and money, there is plenty of good news. Stem cells are already used in clinics with resounding success. Here are the newest medical uses, some still in the last phases of preclinical development and some being tested in humans.

HEALING BAD BLOOD

The tragedy of the Hiroshima and Nagasaki bombings showed how effectively radiation could obliterate the rapidly dividing cells of the marrow. Most radiation victims close to ground zero died within 30 days of exposure. Follow-up research found the only way to save mice from a dose of lethal irradiation was to transplant bone marrow from a healthy donor mouse. The results led others to wonder whether radiation and chemical agents could be used against a disease of rampant cell division, cancer. Their hunch was right, and by 1965 the first cure of childhood leukemia by a bone marrow transplant was announced.[10] The researchers didn't know it at the time, but the marrow's rescue worker was the hematopoietic stem cell or HSC.

There are two basic kinds of bone marrow transplants. Extracting the patient's own healthy cells from the marrow, storing them, and putting them back later is called an autologous transplant. After the marrow is removed, doses of chemotherapy destroy both the cancerous cells and the bone marrow. The cells are then reintroduced to repopulate the marrow, thereby rescuing the patient. Autologous transplants are usually performed when the marrow is healthy and the cancer is elsewhere in the body. In the case of leukemias and multiple myeloma, the marrow itself is diseased. The marrow must be cleansed of cancer cells before it can be reintroduced. The advantage of autologous transplants is that the cells come from the patient's own body, so there is no rejection. New methods can sort the different cells in the bone marrow from each other—similar in principle to the coin-sorting machines found in supermarkets. Clinical studies using blood stem cell purification techniques have found that patients are significantly less likely to have the cancer return and, as a consequence, live longer lives.[11]

An allogeneic transplant uses bone marrow from a different person to treat the cancer. The donor's marrow is removed with a needle, treated, and filtered. Chemotherapy is administered to the patient, and the donor's marrow transplanted. Even the best cell filtration systems can't prevent the donor's immune cells from being transplanted and then attacking the patient—in essence, a reverse kind of rejection. The sometimes-fatal side effect is called graft-versus-host disease (GVHD). Using immune matching to select the best donor greatly reduces the severity of GVHD and cure rates have soared. The closest match usually comes from a family member, such as a sibling. In these cases, immunosuppressant drugs are required to keep the patient from rejecting the donor's blood and the donor's immune cells from attacking the patient.

Stem cells are now used to treat patients who would otherwise have to rely on bone marrow transplants. Rather than drawing out bone marrow through a needle inserted multiple times into the hipbone, the procedure relies on the stem cells circulating in the donor's blood. To increase the numbers of HSCs for the transplant, two drugs are given to the donor. The first is a genetically engineered hormone called granulocyte colony stimulating factor (G-CSF). G-CSF causes the stem cells to expand their number, leave the bone marrow niche, and enter the circulation. Administering a second drug kills rapidly dividing progenitor cells and increases the number of circulating HSCs. Like other bodily insults, the depletion of downstream cells creates a demand to regenerate the blood, prompting new stem cells to enter circulation. The blood is collected over the course of several days and filtered through a machine that isolates the circulating stem cells. Like a regular bone marrow transplant, chemotherapy or radiation therapy is used to kill the patient's cancerous cells and "empty" the bone marrow. The donor stem cells are transplanted and, if all goes well, travel through the blood to the vacant marrow where they colonize and produce red cells, immune cells, and platelets. Even with the bolstered numbers of stem cells, GVHD remains a complicating factor; rogue immune cells cause up to 20 percent of these procedures to fail.[12] New purification techniques can help this problem, too. Removing the immune cells from the donor's HSCs can lessen the chance of GVHD.

Umbilical cord blood has emerged as a new source for transplanting blood stem cells to treat some malignant and nonmalignant blood diseases. Cord blood has only a few primitive blood stem cells because of the small volume of blood found inside—a disadvantage when transplant success is tied to the number of cells infused. The small quantity means that such transplants are suitable only for children or small adults. Nevertheless, using cord blood has advantages. Tests of umbilical cord blood show that its stem cells are highly potent and very active, which means they generate more new blood cells in the bone marrow than their hematopoietic stem cell cousins. Because the immune cells in cord blood are quite immature, an exact match is less important than using stem cells from an older donor. The incoming white blood cells are less likely to attack the patient, resulting in a lower incidence of GVHD. This increases the number of acceptable donors. To boost the numbers needed, mixed cord blood from several donors has been used with good success. As with other rare adult stem cells, the biggest barrier to using cord blood is their limited number and lack of methods to expand cultures to large enough quantities. Stem cell companies are working on methods to multiply cord blood stem cells so they can be used in adult patients.

Many parents donate their child's cord blood for public use. Like bone marrow registries, public banks need a wide variety of cord blood types in order to match donors with recipients. Parents with one sick child (or close relative who is sick) can bank the cord blood of a subsequent healthy child. Cord blood saved from healthy siblings has proven useful for helping children with genetic blood diseases such as sickle cell anemia, thalassemia, and leukemia. If the first child is affected with the disease, the cord blood from a healthy second child can be used as a transplant. At one such bank, Children's Hospital Research Center based in Oakland, California, 46 out of 55 children with blood disease were cured using sibling cord blood-matched transplants.[13]

Should parents of a healthy newborn bank cord blood stem cells for possible use as future therapy for that same child? For fees and dues that run into thousands of dollars, profit-seeking cord blood companies encourage parents to store their baby's blood. The service is designed as a kind of biological insurance against future infirmities, such as leukemia, or for future use of the cells in regenerative medicine. If the family can afford the fees, a future cure for an unseen disease could well be worth the investment. There is some evidence that cord blood stem cells can make muscle and bone, but there is little evidence that they transdifferentiate. Cord blood may help patients with heart disease, too. In any case, it is likely that cord blood stem cells will be first to successfully treat anemias and other childhood blood disorders. For other diseases, families that want to bank their baby's blood will have to weigh the costs of keeping the blood against the likelihood that cures for diseases using cord blood stem cells will be found.

PROTECTION FROM REJECTION

How do immune cells "learn" to protect us? Why do immune cells betray us in diseases like rheumatoid arthritis and diabetes? The answers are found during early development. As our immune system matures, stem cells in the bone marrow develop into antibody-producing B cells, and progenitor cells move into the thymus, where they develop into T cells that protect us against foreign invaders. During fetal development, the immune system learns to distinguish between body cells that are "self" and foreign invaders that are "not self." What remains is a protective apparatus with astonishing flexibility.

Autoimmune diseases occur when the sufferer's body is attacked by its own white blood cells. They destroy cartilage (rheumatoid arthritis), nerves (multiple sclerosis) and organs (juvenile diabetes, Crohn's disease, and lupus). The current treatments only ameliorate the symptoms or slow the disease.

Transplants using embryonic stem cells could cure autoimmune disease, which is actually two problems: (1) a faulty immune system and (2) complications that arise from organs attacked by the immune system's white blood cells. Most of the time, transplants from one person to another cause a rejection of the tissue because it is recognized as "not self." In terms of curing autoimmune disease, the solution sounds too good to be true: a healthy immune system from an unrelated donor can replace a faulty one and, as a result, won't attack the donor's organs or tissues. Stem cell transplantation experiments between mice have proved the concept. At Stanford University, Judy Shizuru and Irving Weissman treated one mouse with radiation, but at a low enough dose that didn't completely destroy the bone marrow.[14] Before transplanting blood stem cells from an unrelated mouse, a purification step eliminated any stray immune cells from the donor that could attack the irradiated mouse and cause GVHD. The incoming stem cells grafted into the marrow, and the mouse recovered and began to manufacture two genetically distinct populations of immune cells—the original ones and the new variety from the donor. The immune cells that develop from the stem cells "learn" to recognize the recipient's cells as "self." After a period of time, the donor's stem cells take over the marrow and become the primary blood-forming system. Amazingly, these "blood chimeras" have a lifelong tolerance to *any* tissue transplanted from the donor mouse.

Human blood chimeras may prove to be the answer for genetically incompatible organ transplants of patients. A donor could supply a patient with a new, compatible immune system and an organ with little or no rejection. For more complicated autoimmune diseases, the brass ring could come from human embryonic stem cells(hESCs). Consider type I or juvenile diabetes, an autoimmune disorder introduced in Chapter 6. Diabetes is ruinous, destroying the insulin-producing β-cell of the pancreas. Complications from this type of diabetes include heart disease, kidney failure, nerve damage, and blindness. In order to cure type I diabetes, the faulty immune system *and* the damaged β-cells must be replaced. Using a line of hESCs made from the patient's own cells won't solve the problem, because immune cells made from the line will have the same defect.

A hypothetical multistep solution using a donor line of embryonic stem cells, shown on the next page, could help patients with diabetes and other autoimmune diseases. First, the patient's immune system must be partially destroyed with chemotherapy. This resets the immune system to the time before it incorrectly learned to attack the patient's own cells and organs. Then, a donor line of embryonic stem cells with a good HLA match would be created. The hESC line would be used to make a population of hematopoietic stem cells. After transplanting the HSCs into the patient, new immune and red blood cells are made. Later, new insulin-producing β-cells made from the same embryonic line would be transplanted to help repair the pancreas. The beauty of this method—called co-transplantation—is no rejection, because both the immune cells and the new pancreatic cells come from the same genetic source and are immunologically compatible. The method could work for other autoimmune disorders like multiple sclerosis: a donor line of embryonic cells could reset the blood system and provide new oligodendrocytes, which are the neural cells that make myelin, the nerve-conducting material.

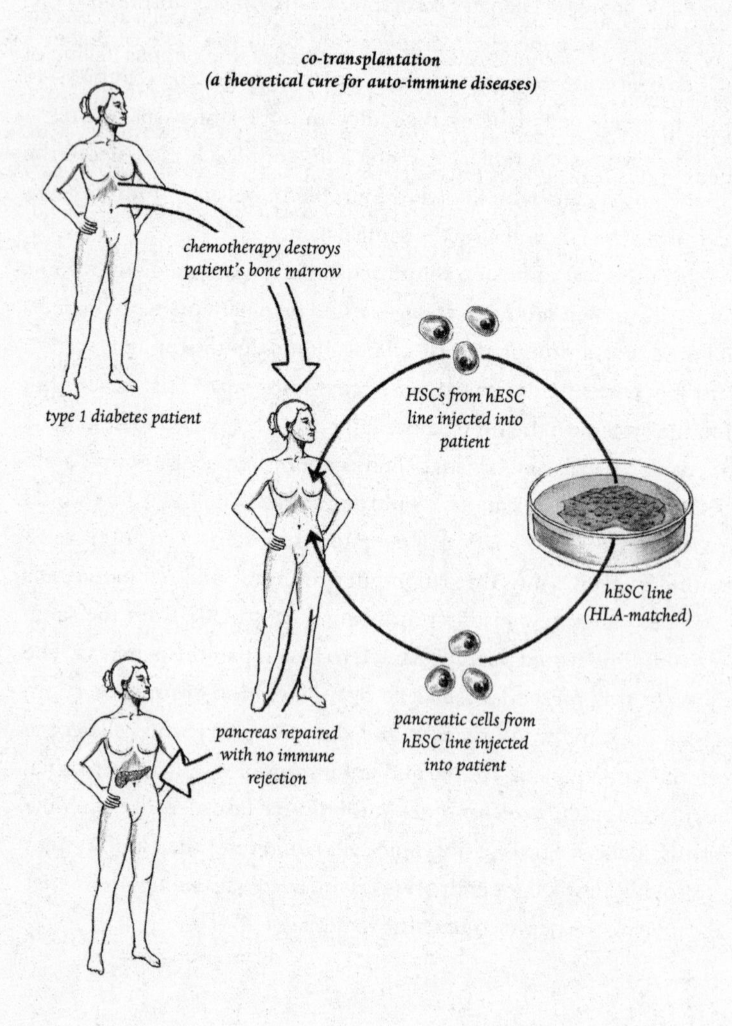
co-transplantation
(a theoretical cure for auto-immune diseases)
chemotherapy destroys
patient's bone marrow
type 1 diabetes patient
HSCs from hESC
line injected into
patient
hESC line
(HLA-matched)
pancreatic cells from
hESC line injected
into patient
pancreas repaired
with no immune
rejection

CELL THERAPIES TO MEND A BROKEN HEART

Cell therapy hopes to treat people with failing organs (such as the heart, lungs, and liver). Clinical trials using adult stem cells are underway. A Maryland company, Osiris Therapeutics, Inc., hopes its line of mesenchymal cells will prevent GVHD after bone marrow transplants. Osiris also uses the cells to regenerate cartilage in injured and arthritic knees and is testing them in patients with heart disease. Building an organ like the heart from scratch is another matter, but stem cells can do the work in damaged areas, much like a road crew repairing a bad stretch of interstate (or inner-state) highway. To be effective therapeutically, stem cells must be made in sufficient quantity and must be shown to repair the organ. Most importantly, the repair must stand the test of time. Using stem cells to treat heart disease is an especially interesting case in point.

Donald Orlic, an associate investigator at the Genetics and Molecular Biology branch of the National Institutes of Health, and his collaborators reported in 2001 that, in heart-injured mice, massive numbers of bone marrow cells repaired the damaged tissue. He interpreted the results as stem cell plasticity: blood stem cells changed into heart cells called cardiomyocytes.[15] The desperate medical need and reasonably safe procedure of injecting heart patients with their own blood (containing hematopoietic stem cells) spurred a clinical trial in Brazil, and by 2003, ten trials in clinics around the world had enrolled hundreds of subjects with end-stage heart disease.[16] Clinicians reported cases where formerly bedridden patients were jogging and climbing stairs after the procedure. In 2003, the British medical journal *Lancet* reported that patients receiving their own bone marrow enriched for stem cells had improved cardiac function and blood flow.[17] The FDA reviewed the Brazilian and European data and moved to approve American trials in hospitals in Texas and Massachusetts in 2004 and in Maryland in 2005.

The foreign trials did show a modest (and statistically significant) improvement in heart function. The percentage volume of blood pumped out of damaged ventricles into the aorta increased up to 8.5 percent.[18] In the meantime, two laboratories working independently at the University of Washington in Seattle and Stanford University could not repeat the Orlic lab experiments. The Stanford group injected highly purified populations of blood stem cells with a genetic tag into the heart muscle of 23 mice.[19] The transplanted cells did not increase the survival rate in mice, did not persist in the heart muscle more than 30 days, and did not produce the signature proteins of heart cells (and instead continued to produce those of blood cells). The cells did, however, improve pumping efficiency slightly. Later, the *Lancet* published the first fully controlled, randomized, and blinded study using unpurified HSCs on heart patients. In 60 patients with myocardial infarction, the average improvement was 6.7 percent.[20] Meanwhile, early reports from the Texas and Boston trials are encouraging. After the procedure, some bed-bound patients are said to be leading normal lives, but their long-term outlook remains unknown.

If sick people get better, then why even debate stem cell plasticity? In the emotionally charged arena of adult and embryonic stem cell research, some worry that the results will cause political winds to shift even further away from acknowledging the benefits of hESCs. To doctors treating sick people, the argument about plasticity isn't as important as the clinical result: what matters is whether the patient gets better. It could be that stem cells improve angiogenesis, the formation of blood vessels. In one recent instance, Advance Cell Technologies' Robert Lanza and his collaborators found that transplanting blood precursor cells found in developing embryos into the damaged hearts of mice repaired nearly 40 percent of the tissue within a month.[21] Lanza's group found that not only did the cells rebuild the heart, the transplants helped the repaired tissue form a new system of arteries and capillaries.[22] Or some other collateral effect might be at work. Because pure stem cells are not used, the effect could be due to something carried along in the mix of cells and liquid. The HSCs could be fusing with cardiac cells, and the union could promote healing. But fused cells do not divide and so can't contribute to heart repair beyond their limited lifespan.

Some laboratory scientists say the clinicians moved too swiftly. Transplanting mixed populations of cells, they say, leaves the mechanisms of the therapy locked in a black box. Harvard's Amy Wagers worries about declaring victory too early. "If we consider a 6 percent improvement in cardiac function a success, then we've left behind an opportunity to understand why this is happening and aim for a 60 percent improvement."[23] Stem cell hunter Kenneth Chien wrote in the April 8, 2004 issue of *Nature*, "We should be wary of prematurely pushing laboratory research into clinical practice."[24] The Harvard scientist went on to say, "Now is the time to search for the presence of naturally occurring heart progenitor cells." About a year later, he found them. In February 2005, Chien discovered a self-renewing population of stem cells in the hearts of adult rats.[25]

Clinicians point out that all the studies were done safely and that some patients fared better after the trials. Joshua Hare is a Johns Hopkins University cardiologist and one of the directors of the university's Institute for Cell Engineering. Hare is using mesenchymal stem cells to treat his heart patients in the Baltimore, Maryland, clinical study. He points out that millions of dollars and years of effort can be spent trying to pin down the minutiae of stem cell science when patients need help immediately. "Basic scientists say it is premature to do the trials when the mechanisms are unknown," says Hare. "I say it's unethical to wait. We won't fully understand the mechanism until we do the human studies. That's what clinical research is all about!"[26]

BIOCONSTRUCTION ZONE: TISSUE ENGINEERING

Stem cells are quietly revolutionizing a high-tech medical field that unites engineering, materials science, and cell biology. The two-decade-old field of tissue engineering relies on the body's own repair mechanisms. A "scaffold" made of nylon or a biomaterial like collagen is transplanted into the body. Cells use the scaffold as support, replacing it over time with natural three-dimensional tissue. Scaffolds can be used with transplanted cells, too. Like the cells in bone marrow transplants, the introduced cells are either from the patient or from a donor.

The success of tissue regeneration varies from organ to organ. The lung, like the heart, has little regenerative oomph. Yet when stem cells from the human lung are purified and put onto synthetic polymer sheets, they form smooth and shiny pulmonary tissue. Also mouse lung tissues made from stem cells develop into respiratory structures after transplantation.[27] Pancreatic tissue hasn't been made, although insulin-producing cell systems are under intense study. The liver can regenerate easily in the body, but a reliable method of growing cells in the lab has not yet been found. Rudimentary skin grafts made using collagen and cell suspensions have helped burn patients for more than 20 years, but regenerating fully functioning skin that does not scar is still a distant goal. For most organs, the secrets lie in the pathways of stem cell differentiation. If science can discover the genes and proteins involved in every step of tissue and organ formation, it may be possible to precisely manufacture the necessary body parts.

Human embryonic stem cells can be used to make beating heart cells, but that is just a start: to be useful as a tissue graft, they must grow into organized structures. Seeding a scaffold made from a thin patch of biomaterial with heart muscle cells begins the regeneration process. Once such structures are sewed into mice, heart cells that secrete matrix-forming proteins infiltrate the patch. This stimulates the construction of blood vessels and, eventually, new myocardial tissue and muscles develop. The heartbeat stretches and strengthens the new tissue as the scaffold slowly degrades.

Making bits of the body's plumbing is proving easier. Like any transplant, the proof of the pudding is whether these stem cell-directed structures can function fully and survive for the lifetime of the patient. Several labs are focusing on forming blood vessels. Another group has devised a way to make a trachea, the thin tube that carries air to the lungs out of cells found in the nasal cavity. MIT researchers are teasing hESCs into cells that line the blood vessels.[28] They seed them on scaffolds, sprinkle in growth compounds, and a few days later watch networks of small tubes traverse through the matrix. When transplanted into a mouse, the manufactured vessels incorporate into the mouse's vasculature.

THE ORGAN MAKERS

The future of stem cell research has also come to the Southeastern United States. Research Triangle Park, North Carolina, is a busy amalgam of biology, technology, and commerce. Wake Forest is just one of several universities that ring the 7,000 acre park, and nestled among the rolling hills are dozens of companies dedicated to research, drug discovery, and healthcare services. Research Triangle Park is billed as the total package, where good ideas can flourish into products. Inside the medical school over an acre of glistening laboratory space is crammed with sleek laboratory benches, special microscopes, and incubators. In one brightly lit room, tall cylinders of clear liquid bubble and boil, and grow *entire organs*.

What does it take to lure one of the world's best tissue engineers away from the heady environment of Harvard University to the tobacco country of North Carolina? A top research university like Harvard, with its medical center and affiliated teaching hospitals, has thousands of faculty members, employees, and graduate and medical students, millions of square feet of space, and billion dollar budgets rivaling many big corporations.

The answer is, quite a bit, actually. It took a multimillion-dollar recruitment package and the promise of a new institute in regenerative medicine for Wake Forest University, with its small but highly regarded medical school, to lure the slightly built and intense scientist away from Harvard's ivy halls. Anthony Atala, 46, belongs to a breed of restless surgeons found in the country's top teaching hospitals. He's confident that what he's created will help the people he treats. "I have one goal," the soft-spoken doctor of urology states flatly. "To cure the patient."[29] Atala is among a growing cadre of physician-scientists who combine their knowledge of human anatomy with an understanding of engineering, electronics, drug development, and computing to devise widgets that will propel us into a buzzing and whirring old age.

The Southeast's dedication to technology is another reason Atala and his team of scientists made the pilgrimage from Cambridge. "Everything we need is here. I can recruit the best scientists to an institution that encourages inventiveness. The local and state governments provide resources and land, and we have capital markets that invest in small companies," he says. In order to make tissues and organs to strict standards, the university gave him funds to build a government-certified facility to manufacture organs near the institute. The institute also has a venture fund, managed by Wake Forest, that invests in the most promising medical discoveries. Atala hopes that once his new ideas hit the market, the profits from products and public offerings will flow back and fund more research—a virtuous cycle.

It's an interesting combination; the razzle-dazzle high finance of biotechnology linked to a man who believes that slow and steady will win the race. After years of experimentation, in 1999 Atala created the first human organ—a bladder—using tissue engineering. The bladder looked pretty normal—oblong, hollow, with a narrow top and bottom—and was complete with blood vessels, nerves, musculature, and openings in the right places. Other organs followed. Using a sample of human tissue, he's been able to make cell suspensions that form uteruses, vaginas, and large blood vessels. Believe it or not, Atala's group has constructed a fully functioning rabbit penis. Randy rabbits with newly engineered members chase their female counterparts around the cage barely four weeks after surgery. His laboratory equipment is mostly custom built. Fish-tank-sized structures connected to transparent hoses are controlled by a blizzard of dials and gauges. A handful of his new organs have already been transplanted into humans. Atala doesn't rush things—he's still observing patients with manufactured bladders four years later and aims to try larger clinical trials only when he's certain he's found the safest and most efficient procedures.

Ten years ago, they said human organs couldn't be built. Now the challenge is unraveling the knotty problem of solid organs, like the liver, pancreas, heart, and lungs. Using renal tissue from cows and a spongy matrix no bigger than a 50-cent piece, the institute has made miniature kidneys that filter blood and eliminate straw-colored fluid. The surprising thing about many of these successes is that they don't rely on a pure source of stem cells. As long as the right kind of stem cell is in the mix of tissue, the brew of growth factors and scaffolds do the rest.

"Most of our research has a stem cell focus, both adult and embryonic," he maintains. "Some people are under the impression that regenerative medicine is this science-fiction drama with dozens of ready-made organs hanging in refrigerators, waiting for patients. That's nonsense." He waves his arm toward a collection of scientists and students huddled over an organ culture machine. "The reason that we've assembled an international group of physicians, engineers, and biologists is that one technology doesn't solve all problems. We have to change our expectations about regenerative medicine. The absolute first priority is the patient—what's best for the patient?"

To illustrate, Atala recites a litany of broken and diseased human parts and the approach for each. "If a heart has an infarct [or damage]," he says, "engineering new tissue from embryonic stem cells is best, because trying to biopsy a bit of the heart in order to isolate cells for tissue engineering endangers the patient." The same goes for especially sensitive organs like the pancreas, where the slightest injury provokes pancreatitis, an inflammation that can destroy the organ. Atala flashes his frustration with conventional surgical wisdom. "Why is it that surgeons think that if a piece of your heart gives out, you have to change the whole heart? You don't! Our organs have tremendous reserves. When someone comes to a doctor with heart pains or kidney trouble, it's because 90 percent of the organ has failed. You don't need much repair to get back to a normal lifestyle. And a stem cell patch may be the best approach."

When asked about whether making customized organs and tissues patient-by-patient will be cost effective, Atala replies, "You can't argue with autologous." His point is that making organs from a patient's own cells is the best way to go, regardless of cost. He claims he can build a hollow organ in just five weeks: four weeks to expand the number of cells and one week to seed and build the "construct," the three-dimensional structure that becomes the organ. The advantage is no tissue rejection. Once again, Atala cuts to the chase: "Immunosuppressant drugs are nasty things. I think that people who suggest that we can control rejection with better HLA matches haven't spent much time at the bedside of someone on prednisone. There are too many immune genes that make us different, and more are discovered every year. I don't believe that we'll solve immune rejection in my lifetime, or perhaps ever."

So to get where he wants to go, Atala is focusing on how nature does it and what's best for the patient.

REPAIRING NERVES

The "before" videotape shows a rat with a recently damaged spinal cord. Staggering inside a circular Plexiglas container, the animal drags its hindquarters along the runway, its tail trailing limply behind. The next segment shows the same rat after an injection of oligodendrocytes made from a line of embryonic stem cells. It sniffs the air and stands on its rear legs. Dropping down, it takes a lap around the cage with only a slight suggestion of a limp—once disabled, now cured. The revived rat is compelling visual evidence that Professor Hans Kierstead uses to wow audiences of scientists, patients, and news reporters. The University of Irvine neuroscientist is backed by the Christopher Reeve Foundation and the stem cell company Geron. His recently published research on rats has caused a stir among patient activists and physicians who are impatient for human testing to begin.[30] He takes embryonic stem cells and differentiates them into pure colonies of oligodendrocytes, the neural cells that form the conductive material myelin. The cells are then injected seven days after the injury, and the rats began to walk properly within two months of the treatment. Timing is crucial: when Keirstead waits until 10 months after the injury, motor movements do not return. The rat's trauma doesn't sever the nerves—leaving the basic wiring essentially intact—and it's not clear how effective the procedure will be for injuries involving severed nerves or scar tissue. Geron hopes to clinically test the safety of their therapy in as early as 2006, during the routine surgery that follows an accident. Repairing damaged human nerves can be a major medical victory—the success would ripple far beyond spinal injury to demyelinating diseases, such as multiple sclerosis.

Chapter 3 detailed the chain of developmental events that lead to the most complex network of cells in the body, the nervous system. When it comes to understanding how nerve cells come to be, Anders Bjorklund, professor and chief of the Wallenberg Neuroscience Center at Lund University, Sweden, says that "compared to our understanding of the blood stem cell system, we are at least a couple of decades behind."[31] Bjorklund, with support from the Michael J. Fox Foundation, pursues clinical research in Parkinson's disease, where loss of cells that produce dopamine causes neurons to fire out of control; this results in loss of motor movements.

At this writing, no human clinical trials are using stem cells for Parkinson's, but there is a history of treating Parkinson's patients using neural cells from aborted fetal tissue. Bjorklund's ground breaking work in the late 1980s transplanted six-to eight-week-old neural fetal cells into the brains of humans and proved that cell therapy could actually work.[32] In the majority of patients, the injections improved motor function. Follow-up studies have been less encouraging. A clinical trial in 1999 at Columbia University and the University of Colorado had mixed results, helping younger patients but offering no benefit to patients over 60.[33] In 2001, the same physicians did a follow-up study, but this time tremors in 6 out of 20 patients receiving fetal cells became worse.[34] The results worried many who feared that cell therapy could be more damaging than therapeutic, especially when treating brain disease. As a result, preclinical work with stem cells is moving slowly through animal testing. Chapter 6 describes how embryonic stem cell therapy can improve Parkinson's-like symptoms in rodents and monkeys.

Slow-progressing brain diseases are heartbreaking, but there is an affliction that is even worse: Batten's disease. In Batten's patients, a defective enzyme keeps cells from degrading lipoproteins, fatty packages of cholesterol that travel through the bloodstream. Lipoproteins accumulate in the cell's cytoplasm to the point where they destroy neurons, retinal cells, and brain cells. The afflicted individuals go through rapid stages of blindness, ataxia, dementia, and finally, death. Batten's disease is fatal and affects only children. In the United States alone, 10,000 infants will appear healthy until their first birthday. By age 3 they will be dead.

Scientists at the biotech company Stem Cells, Inc., believe they have an answer for Batten's patients. They hope their discovery will also lead to treatments for other lysozomal storage disorders like Gaucher and Tay Sach's disease. The researchers used stem cells found in the human brain to perfect a procedure that multiplies them into clusters of multipotent neural stem cells. After human cells are transplanted into a mouse with a disease that mimics Batten's, the stem cells travel to the damaged areas and begin to secrete normal enzymes, slowing the progress of the mouse's symptoms. Ann Tsukamoto, Stem Cell, Inc.'s vice president of Research and Development, is encouraged by the results. "The fact that we see increasing levels of enzyme production over time is very positive," she says. "It means that transplanted cells are renewing, multiplying, and might be replacing dead cells with normal neurons. We're hopeful that a single transplant may be enough to treat these patients."[35] The company will start human clinical trials in 2006.

STEM CELLS AS TOOLS

The experts agree that immortal embryonic lines will become one of biology's most powerful tools. The Australian embryologist and stem cell biologist Alan Trounson, a pioneer of *in vitro* fertilization, believes that "studying disease with stem cells is incredibly important for research. We need to develop embryonic cell lines from patients who've got muscular dystrophy, Alzheimer's disease, and cystic fibrosis. That way we can develop drugs that actually block the disease from occurring."[36] James Thomson agrees. "Human embryonic stem cell research will be developed more as a research tool than for transplanting engineered cells and tissues. I mean, think about disease for a minute," he says. "You don't want to do anything so crude as replacing those cells once they have died. You want to stop the disease from happening in the first place! If you had a reliable supply of neuronal cells, for example, you could study them to understand exactly how Alzheimer's disease causes them to die."[37]

Discovering drugs is an important application of hESC technology. Potential drugs made of chemical or biological compounds can be tested in cultures of pure populations of cells that are specifically related to or affected by the disease. For example, the dopamine-producing neurons implicated in Parkinson's disease might be made from hESC lines and stored in quantity. Treating the neurons and measuring their response would quickly sort out which chemicals work best. Thousands of potential drugs tested in this fashion would speed up drug discovery. Existing pharmaceuticals could be refined and improved in the same fashion.

Gene therapy is a relatively recent and highly experimental approach to treating disease. Although most drugs are manufactured outside the body, gene therapy takes a different approach: a gene is delivered into the affected cells in the body, where it produces a protein that acts as a therapeutic agent. The potential success depends not only on the gene's delivery into the appropriate cells, but also on the gene's ability to function properly. Both requirements pose considerable technical challenges. Noninfectious viruses are used to deliver the gene, just like ordinary viruses infect cells. Unfortunately, this method is imprecise and limited to the specific types of cells the virus can infect. If the proteins aren't produced efficiently or the transformed cells eventually die of old age, then repeated rounds of therapy are needed.

Gene therapy can be improved by using stem cells. Because stem cells self-renew, they can reduce the need for repeated rounds of therapy. Blood-forming stem cells are especially good choices for delivering drugs because they are easily removed from—and reintroduced into—the body, and once in the body, they home in on certain organs and structures such as marrow, spleen, and thymus. Dozens of human clinical trials have used HSCs to deliver therapeutic agents such as interferon to patients suffering from blood and solid-tumor cancers (as opposed to cancers of the blood), anemias, and immune diseases such as SCID and HIV.[38] In some cases the results have been promising, extending the lives of terminally ill patients. Cell-to-cell fusion—one of the phenomena behind apparent stem cell plasticity—might also be a way to deliver a therapeutic gene. If the disease is due to a missing or defective gene in the liver, an engineered blood stem cell might fuse with liver cells and produce the needed protein. However, fusion is a rare event, so delivering enough protein to repair the organ may be a problem.

Neuroscientist Anders Bjorklund believes that a combination of gene and stem cell therapy holds the key to correcting brain dysfunction. He's set his sights on a mutation in a gene implicated in Parkinson's called Nurr1. If a corrected copy of Nurr1 can be delivered to patients via stem cells, he believes it will slow or stop Parkinson's progression. The idea is to swap a corrected copy of the defective gene into an uncommitted neural cell. Many such engineered cells could be injected directly into the brain. If the cells took hold, they would manufacture the missing protein. Bjorklund adds that cells that promote brain healing and self-repair could be injected. The big hurdle here is navigating the pathways of cell differentiation. According to Bjorkund, "One of our dilemmas is that we don't always know what is, and what is not, a nervous system stem cell."[39]

Experiments using other kinds of stem cells to carry therapeutic cargo are underway. High-powered mesenchymal cells carrying the cancer-fighting gene for interferon doubled the survival rates for transgenic mice ridden with human tumors. The cells homed in on tumors no matter their location, suggesting a way to treat human cancers that have already spread.[40] Neural stem cells carrying antitumor agents such as interleukin 2 also show promise. In mice with brain tumors, the stem cells produce and release interleukin 2, which then stimulates white blood cells to enter and kill the cancer cells. Other neural stem cells deliver the drug right to the cancer cells themselves, ensuring a good kill rate.

Stem cells are being used to diagnose disease. At a growing number of *in vitro* fertilization clinics, single-gene defects that cause Huntington's, Tay Sachs, sickle cell anemia, cystic fibrosis, and dozens of other disorders are being detected via an embryo-sampling technique called preimplantation genetic diagnosis, or PGD. Four days after fertilization, while still in a laboratory dish, an eight-cell embryo is grasped gently by light suction and a single cell is removed with a pipette. The embryo recovers with a quick round of cell division. The DNA in the cell is extracted and then tested with a genetic probe for the disease in question. The test ascertains whether the embryo has no disease genes, is a "carrier" with one disease gene and one normal gene, or has both copies of the gene and will therefore develop the disease. Only embryos with no disease genes are chosen for implantation. Parents who carry the gene or who have family histories of the disease can use PGD to avoid having an affected child. A separate analysis identifies a group of different diseases caused by the wrong chromosome number, such as Down's syndrome, or abnormalities that lead to miscarriage. PGD has emerged as a tool for parents whose only other option would be to test abnormalities during fetal development. In most cases, PGD enables the family to avoid the difficult decision of whether or not to end a late stage pregnancy. The procedure is expensive, running $5,000 for both sets of tests.

Just as advances in reproductive biology helped embryologists derive the first hESC lines, PGD can help study disease. Rather than discarding donated embryos that test positive for defects, a clinic at the Reproductive Genetics Institute in Chicago has developed over 30 hESC lines by transferring the defective nucleus into enucleated eggs.[41] These stem cell lines, each with a different genetic disease, are now available to researchers who can use them as an *in vitro* model. Observing how these cells behave compared to normal cells will help identify how certain diseases begin, progress, and affect healthy tissue. Not only are the disease-causing genes and their proteins identified, this also opens up possibilities for designing drugs that reverse or treat the problem.

The story of Molly Nash illustrates how stem cell tools and therapies can work together to save lives. The Colorado child was born with Fanconi's anemia, a genetic blood disease with an especially poor prognosis. Most patients rarely reach adulthood and die of leukemia. A bone marrow transplant from a healthy sibling with a matched HLA or immune profile can cure the disease, but Molly was an only child and her parents—both carriers of the deadly gene—were fearful of having another child with the disease. They used *in vitro* fertilization, pre-implantation diagnosis and a cord blood transplant in an attempt to save their child. PGD was used to screen 24 embryos made in the laboratory. One embryo was disease-free and matched Molly's immune profile. The blastocyst was implanted and, nine months later, her sibling, named Adam, was born. The stem cells from Adam's umbilical cord were given to Molly and, today, she is 11 years old and free from disease.

The diversity of stem cell treatments reflects the diversity of stem cell breeds. Lines of hESCs may become the preferred source of cells used to treat patients. Presently, tissue engineering, bone marrow transplants, and the results from early clinical trials confirm the utility of adult stem cells taken from the body. Along ethical dimensions, the two kinds of cells are very different, and those differences deserve a closer look.

8

The Great Moral Divide

Nobody, I imagine, will credit me with a desire to limit the empire of physical science, but I really feel bound to confess that a great many very familiar and, at the same time, extremely important phenomena lie quite beyond its legitimate limits.

THOMAS HUXLEY[1]

"The newly fertilized egg, a corpuscle one two-hundredth of an inch in diameter, is not a human being. It is a set of instructions set adrift into the cavity of the womb," wrote E. O. Wilson, a Pulitzer-prize winning zoologist.[2] But where E. O. Wilson seems unperturbed, Charles Krauthammer, M.D., a syndicated columnist and member of the President's Council on Bioethics (PCBE), a 17-member group charged with advising the president on the ethics of biological innovation, is deeply concerned. "We will, slowly and by increments, have gone from stem cells to embryo farms to factories with fetuses hanging (metaphorically) on meat hooks waiting to be cut open and used by the already born."[3]

The two quotes reflect opposite poles of an emotionally charged debate about the use of human embryonic stem cells (hESCs) for research and medicine. The rhetoric often reveals from which side of the debate the person speaks. Supporters of hESC research sometimes describe the human embryo as nothing more than a ball of cells; opponents mention the embryo as if it were a newborn baby. Others portray the blastocyst as something in between: neither a clump of cells nor a person. All who consider the question of using embryos for research share one thing: passion. Embryonic stem cells touch us deeply, not just because they might cure disease. It is because forms of human life are at stake—living embryos and living persons.

Many ethical questions touch hESC research. What are our moral obligations to the sick among us who could benefit from embryonic stem cell research? What roles do family, religion, and society play, and how do they inform our opinions and decisions? How do we properly explain the benefits and risks of using embryos and cells to the individuals and parents who donate them? Who should benefit from the first therapies, and how will we pay for them? Tackling these questions could fill an entire book. But the ethical question on most people's minds (including people in the federal government) is whether a four- to six-day-old human embryo—the blastocyst—should be used to make lines of hESCs. Adult stem cells taken from a consenting person's body don't raise the same ethical concerns. Adult stem cells made from embryonic stem cells do.

A much more incendiary issue is whether somatic cell nuclear transfer (SCNT) should be used to make cloned humans for medical or other purposes. Here, the ethical positions are unanimous: both sides of the debate overwhelmingly condemn human reproductive cloning. Making a cloned human would require surmounting all the difficulties inherent in nuclear transfer, many of which may never be fully understood. Attempting to overcome these problems in order to achieve a normal birth or to use embryos to make fetuses or human beings for spare parts would amount to the worst forms of illegal and unethical human experimentation. Our moral codes are designed to protect people in our pursuit of knowledge, and humans or fetuses cloned for research purposes fall squarely and unambiguously into this category. Despite the universal agreement on prohibiting human cloning, some, like Charles Krauthammer, worry that if we use embryos for research, unscrupulous persons will use them to create more developed forms of life, including humans. Because of this, they argue, it is best to err on the side of safety and not use embryos at all.

PHILOSOPHY MEETS THE NEW BIOLOGY

Philosophers, theologians, and entire societies are wrestling with these questions, so much so that a new field, bioethics, has emerged to study the advances of medicine and technology and their impact on humankind. Bioethics is a child of the field of medical ethics, the qualities embodied by practitioners of medicine, who are enjoined to heal and not to harm; to (if asked) refuse to extinguish life; and to strictly keep the confidence of their patients. But, 40 years ago, the "new biology" entered the arena and, with it, an avalanche of discovery. Scientists could insert foreign genes into living things and transplant human organs. No longer do we think only about the pros and cons of traditional medical care; we now face the ability of science to change our human future—and change it radically.

Bioethics is still a young field, and the rapid pace of stem cell biology is just one area challenging ethics to stay in step. In academia, dozens of programs in bioethics, most within medical schools, grapple with a wide range of rapidly evolving issues. One of the early thinkers in the field, Albert Jonsen, a specialist in medical ethics and former presidential advisor, writes, "Only half of bioethics counts as an ordinary academic discipline: the half that has original and borrowed theory, principles, and methods. But only part of bioethics lies within the academy, where scholars worry about whether they have a discipline to teach and promote. The other half of bioethics is the public discourse: people of all sorts and professions talking and arguing about bioethical questions."[4]

Jonsen is right about public discourse. Our moral universe is changing and challenging us, impinged upon and steered by ethics debates.

Over time classic ethics endeavored to describe our world and the best prescriptions for living in it. The results often translate into laws and public policy—the voice of governments and institutions. When it comes to stem cells, the positions taken by scholars and philosophers are strewn across the map. It is worth exploring the views of a handful of prominent bioethicists—including members of President George W. Bush's ethics council—to see how the discourse on stem cells is shaping up.

THE MORAL STATUS OF THE EMBRYO

Of all the moral dimensions of stem cell research, the human embryo looms largest. Recall that viable embryos may be created two ways: naturally, through the process of conception, or artificially, through either nuclear transfer or *in vitro* fertilization (IVF). In the latter case, removing the inner cell mass (ICM) in order to produce stem cells destroys the embryo. The cells may survive indefinitely, but the embryo is gone. Fertilized eggs kept in a freezer or blastocysts grown in a lab will never develop into a fetus or human because the outer layer of cells, the trophoblast, is missing. Normal development and a live birth require a viable embryo's successful implantation into the uterus, where it will grow to term.

What should we do with leftover embryos from IVF procedures? Are we justified in using them for the purposes of research? Should we create embryos using nuclear transfer so that we may use them for therapies? Do we treat embryos the same as a newborn or a child or even as an adult? The moral status of an embryo—whether we consider it a mere object, a human being, or somewhere in between—is the subject of thousands of pages of opinion, essays, and research. A rough sorting divides the issue into two camps. One group believes that embryos deserve protection and should not be used for research. The other group believes that embryos can be used and embryonic

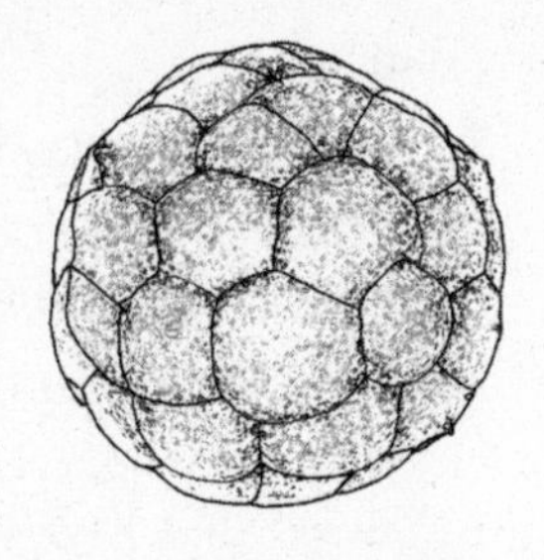

stem cell research should proceed. The divining rod prompting each stance is the moral weight given to the embryo.

The blastocyst depicted is shown 330 times larger than life size. If we place a culture dish under a microscope and peer at the embryo at this stage, should we protect it from harm? If not, should it be used to save others from disability, disease, and death? The debate pivots around these two questions.

PROTECT THE EMBRYO

The answers depend, in part, on whether we believe the embryo has a soul or is a person; in essence, whether it is a human being. If so, then we should protect it. Presidential council members Robert P. George and Alfonso Gómez-Lobo—the first a political scientist and the second a philosopher—are clear about where they stand. They write, "A human embryo is a whole living member of the species *Homo sapiens* in the earliest stage of his or her natural development.... The human embryo, then, is a whole (though immature) and distinct human organism—a human being."[5] Their position is based on the fact that human development proceeds along a continuum. Each of us was once a one-celled zygote, embryo, fetus, and an adult. To George and Gómez-Lobo, each stage represents a different version of the same person. If the zygote has the "natural capacity" to become a person, they argue, then it's an either/or matter: "a 'thing' either is or is not a whole human being."

Conservative Christian thinkers agree. Gilbert Meilaender, a Lutheran theologian and member of the president's council, considers the embryo "the weakest and least advantaged of our fellow human beings."[6] Meilaender believes that the embryo's mere presence is enough to make it a person, though we may be barely aware of it during the early stages of a pregnancy or when it is stored for use at a reproductive clinic. He warns that some of his fellow philosophers define the concept of "person" too narrowly, "trying to find some capacity—perhaps self-awareness, reasoning power, or sense of oneself as having a history—that marks the point at which human beings become persons (or cease to be persons)."[7] The Valparaiso theologian might disagree with the English philosopher John Locke (1632–1704), who described a person as "a thinking, intelligent being, that has reason and reflection, that can consider itself as itself, the same thinking thing, in different times and places."[8] The traits Locke describes can be broadly interpreted: humans who have lost or never acquired these abilities may not be persons; animals with these traits could be construed as persons.

"I think it is accurate to state that a person is simply a 'someone who'—a someone who has a history," Meilaender says. "That history begins before we have some of the highest human capacities and, for many of us, it may continue after we have lost those capacities. But we remain the same person throughout the trajectory of our life." Meilaender's concept of personal history—and *personhood*—thus begins before we are conscious of it and continues after we have lost consciousness. Acknowledging this history gives the embryo the same amount of respect that we would give a fully conscious adult human being.

Some philosophers use a different rationale to argue for the protection of embryos. They say that our uncertainties about embryos are enough to give us pause. Although council advisors George and Gómez-Lobo characterize human development as a seamless series of the "same person," those who aren't sure about that claim say that it is impossible to pinpoint when the metaphysical characteristics of soul and personhood begin. Perhaps the blastocyst has a scintilla of personhood at four days—how can we really know? If an embryo is *possibly* a person, they contend, then it follows that destroying an embryo could possibly mean killing a person. Given this doubt, we should leave it alone. The Christian ethicist Robert Song believes that using embryos for research and medicine requires a "burden of proof" that they are *not* persons. "The merely possible personhood of the embryo may *seem* abstract or theoretical in comparison with the concrete hopes for clinical treatments." But it is proof enough, he says. "[F]or all one knows, they are persons, and should be treated as persons."[9]

A similar argument rests on the fact that a one-celled embryo *will* become a person, circumstances permitting. It uses a concept rich with meaning in stem cell biology: potential. A fertilized egg has the potential to become a fully realized person and, when an embryo is destroyed, its potential to become human is destroyed, too. When something dies naturally, as in the case of a fertilized egg that never implants in the uterus, or intentionally, when a less-than-perfect zygote is discarded after an IVF procedure, the occurrence causes us to reflect and wonder what that embryo could have been. President George W. Bush's former chief advisor on bioethics, Leon Kass, puts it this way: "[I] must acknowledge that the human blastocyst is (1) human in origin and (2) *potentially* a mature human being, if all goes well."[10]

The official positions of conservative religions consider the embryo as a human being. The Southern Baptist Convention, noted for their long-standing position on abortion, strenuously opposes human embryo research and calls upon research centers to "cease and desist from research which destroys human embryos, the most vulnerable members of the human community."[11] Both the Orthodox Christian and Methodist churches assert that embryo research is a fundamental violation of human life. The Anglican view, prepared as a brief for the House of Lords in 2002, elevates the moral status of the embryo as sacred, containing "the very beginning of each human being."[12]

The Catholic Church declares that God bestows personhood and a soul at the moment of conception. The official position of the Vatican, however, has shifted over the years. Saint Thomas Aquinas (1225–1274) believed that a human soul was not present at conception, but appeared between 40 and 90 days later.[13] In 1974, the Vatican wrote in its *Declaration on Procured Abortion*, "Respect for human life is called for from the time that the process of generation begins. From the time that the ovum is fertilized, a life is begun which is neither that of the father nor of the mother, it is rather the life of a new human being with his own growth."[14] The church went on to define an embryo as a person in 1987. The instruction, called the *Donum Vitae* (The Gift of Life), explains, "From the time that the ovum is fertilized ...[it] demands the unconditional respect that is morally due to the human being in his bodily and spiritual totality...[and] his rights as a person must be recognized, among which in the first place is the inviolable right of every innocent human being to life."[15]

For the Vatican, the unique combination of genes resulting from meiosis and fertilization is enough to trigger personhood. Using the *Donum Vitae* as its guide, the church condemns the use of embryos for research. In 2000, it wrote that removing the inner cell mass from an embryo "is a gravely immoral act, and consequently is gravely illicit."[16]

USE THE EMBRYO

On the other side of the moral divide are those who believe that embryos are not people and maintain that we can—indeed must—use them to help humankind. Not surprisingly, many scientists, patients, and doctors hold this view. Moderate and left-leaning philosophers and theologians tend to be in this camp and disagree—at times vehemently—with their conservative colleagues on the moral status of the embryo.

If polls are any indication of American religious sentiment, they show a disagreement between those who profess faith and those who institutionalize it. The fracture is particularly notable among mainstream American Protestants and Catholics. In a Harris Poll conducted in August 2004, 73 percent of Catholics voted in favor of embryonic stem cell research; only 11 percent were against it. The margin of

Protestants in favor of research was even larger—8 to 1. Even among the "evangelical" or "born-again" Christians so important to the conservative political right, only one in five voted against using embryos.[17]

Other exceptions to the "bright-line" religious views include the American Presbyterian Church. Its resolution places the respect of sick persons above the respect due the embryo, affirming the use of stem cells "for research that may result in the restoring the health of those suffering from serious illness."[18] A growing number of centrist Catholic theologians don't consider the embryo before two weeks of age an individual human entity. Other moderate Catholics use an ethical middle ground called "proportionate reason," which tolerates a "lesser evil" to bring about a "greater good." Michael Mendiola, a Catholic and professor of Christian ethics in the Graduate Theological Union at Berkeley, California, suggests one version of this compromise: a requirement that researchers consistently seek to move beyond the use of human embryos while allowing their use in the near term.[19]

Lutheran Theologians Ted Peters and Gaymon Bennett use a bioethical principle of beneficence—a moral obligation to act for the benefit of others—to make a case for supporting embryonic stem cell research.[20] They recall the parable of the Good Samaritan, who pursues one goal: to heal a suffering stranger. The duo contrasts positive acts to help others against an instinct to protect the embryo, a position they summarize as "better safe than sorry." Bennett, an affable young religious scholar who studies at the Center for Theology and the Natural Sciences in Berkeley, California, doesn't fit the image of a gray-haired philosopher with a furrowed brow. He says he wants a just world—where people suffer less. "I think about how science can help me bring about that kind of world. But I also consider how science can get in the way of achieving that aim."[21] He notes that for most people, the stem cell debate comes down to three questions: What are our responsibilities for (1) people who are sick and injured, (2) the embryo's protection, and (3) the unseen consequences of the research?

On question (3), Bennett says, "We shouldn't believe 'anything goes.' Ethics argues against doing science purely for science's sake. We need a philosophy that guides science, and we must realize we won't always have a perfect match of scientific and ethical truth."

Even with these cautions in mind, Peters and Bennett say the potential benefits of medicine should come first. Opponents of hESC research maintain that we should uphold human dignity and, as a result, protect the embryo. The two theologians point out that we aren't certain about when an embryo or fetus becomes human. In the meantime, we are certain there is plenty of human suffering, and these people could benefit from discoveries made with embryonic stem cells. Because of this certainty, they argue, we should pursue hESC research.[22]

The Jewish faith's assessment of the early embryo's moral position stands in stark contrast to conservative Christian dogma. Laurie Zoloth is a Jewish philosopher and professor of medical ethics at Northwestern University. Jewish religious law—called *halachah*—uses a defense of justice in all of its deliberations about medicine and health-care. Zoloth describes the mandate *pikauch nefesh*, the duty to save a life and to heal. If healing is mandatory, then embryonic stem cells *must* be used. She explains that with the exceptions of the ill-begotten use of murder, adultery, or idolatry, "If one can save a life, one must save a life; if one can heal, one must heal."[23]

Is breaking apart an embryo in the lab murder? Halachah says no, it is not considered murder, nor is it considered killing. Zoloth uses an example of Judaic case law to illustrate. "Suppose two men are fighting and as a consequence a pregnant woman is injured as an innocent bystander. If she has a miscarriage, we consider the harm has come to the woman—not to the potential child." The rabbinical text treats pregnancy as a continuous process, with the moral status of the embryo, fetus, and newborn gradually increasing along the way. Pregnancy—along with a significant moral imprint—isn't officially recognized until after 40 days. "The old texts describe the developmental period before 40 days 'like water,'" Zoloth says. "It makes practical sense if you think about it. If a pregnancy ended before 40 days, you wouldn't be able to detect a recognizable human form with the naked eye." The same reasoning applies to embryos in a laboratory dish. They don't have true moral status unless they are implanted in a woman and survive until 40 days. Zoloth points out, however, that we should not be careless with embryos and that they should be treated with special consideration and respect.

Jewish religious leaders emphasize the importance of the duty to heal. In a written statement prepared for a government commission on bioethics in 2000, Rabbi Elliot Dorf of the University of Judaism contends, "[I]n light of our divine mandate to seek to maintain life and health, one might even argue that from a Jewish perspective, we have a duty to proceed with that research."[24] In the same session, Yeshiva University's Rabbi Moshe Dovid Tendler extends a warning, "[A] fence that prevents the cure of fatal diseases must not be erected, for then the loss is greater than the benefit. In the Judeo-biblical legislative tradition, a fence that causes pain and suffering is dismantled. Even biblical law is superseded by the duty to save lives."[25]

Interpretations of Islamic law also find support for the use of embryos for research. As in many cultures, Islam's debate about the status of extracorporeal tissue arises from its discussion about cells taken from aborted fetuses for therapies. Opinion and law about embryos and stem cells are sparse, but in a review of both Shi'ite and Sunni interpretations of the Koran, the Islamic scholar Abdulaziz Sachedina mentions that most opinions state that the soul is formed sometime after the early stages of embryogenesis. As a rule, Islam usually accepts abortion before the eightieth day. Like Judaism, Islam permits interventions in nature that will further a greater good for humanity. But Sachedina also notes that there is disagreement among Muslim jurists about whether embryos in IVF clinics have a predefined sanctity, ensuring their protection, or no sanctity, allowing their destruction.[26]

PHILOSOPHICAL ARGUMENTS

Philosophers and ethicists describe the concept of *sentience*, the capacity to feel psychic or physical pleasure or pain. In order to be sentient, an animal or person must be conscious of its existence. Permanently unconscious persons—including those classified neurologically as being in a "persistent vegetative state"—lack sentience. Though we think of them as human, we also acknowledge that they lack an essential quality of "being in the world," that is, relating, speaking, and thinking like other members of society. Such experiences make certain creatures not just a thing, but also a being. In other words, consciousness, whether it is explained behaviorally or neurochemically, is the experience of *being a being*.[27] Sentient beings deserve our respect and demand that we're attentive to the possibility of causing them pain, fear, or discomfort.

Biology isn't adept at defining when a soul becomes present or a when a person emerges during development. However, biology can be helpful when we think about sentience. The threshold of sentience can be estimated in developmental and biochemical terms, when the nervous system has matured. When are the sensations of pleasure and pain and an inkling of consciousness possible? At this juncture, it's best to ask a neuroscientist.

Michael Gazzaniga has spent a career at the edge of brain biology and is another member of the PCBE. As we learned in Chapter 4, the formation of the brain is a *tour de force*, a long process that begins early in development. The neuroscientist disagrees with some of his fellow council members about the status of an embryo, and bases his position on what he understands about brain biology. He points out that the human brain isn't viable until gestation has reached six months—

about the same time a fetus could survive a premature birth in a hospital's neonatal unit.[28] Working backward through time, the brain's first electrical activity is detected around week six and the major divisions of the brain don't form until week four. The primitive streak—an indentation on the embryo that signals the scant beginnings of a nervous system—appears at 15 days.

In contrast to those who consider an embryo the same moral person at different stages, Gazzaniga believes that moral rights of the developing human begin at six months, coincident with its survival outside the womb. Just as Zoloth describes the early stages of embryogenesis "like water," Gazzaniga uses a perceptual (and emotional) cue to imagine brain function at eight weeks, when the head of the fetus begins to look human. He writes, "I am reacting to a sentiment that wells up in me, a perceptual moment that is stark, defining, and real. The brain at [eight weeks] is hardly a brain that could sustain any serious mental life."[29]

There are other, nuanced arguments made by philosophers and commentators who support hESC research. James Wilson, an emeritus professor of public policy at the University of California, Los Angeles, resigned from the PCBE in May 2005. He wrote in the council's 2002 report, "My view is that people endow a thing with humanity when it appears, or even begins to appear, human—that is, when it resembles a human creature. The more an embryo resembles a person, the more claims it exerts on our moral feelings."[30] Others note that had a modern version of "technological ethics" been present a century earlier, we could well be without air travel, space flight, or the benefits derived from other risky areas of science.[31] A number of scholars respond to religious "sanctity of life" views by pointing out that up to half of fertilized eggs never implant in the uterus or, if implanted, fail to develop. If the Vatican contends that life begins at conception, then, the argument goes, how could God create a system for making babies that destroys half of them?[32] Zoloth says, "Of all the things I don't understand about

God's plan, I do understand this: part of the way the world is structured is that every time you become pregnant, you do not produce a living being. I know that one can be a woman of faith and experience a miscarriage and understand that you have not failed."

The assertion that an embryo should not be destroyed because it has the potential to become a person or possibly is a person is countered by a world full of real people, each with a fully realized potential. Professor of ethics and theology James Petersen writes, "How can we let patients who are unmistakably people die to protect embryos that, even if implanted, may or may not turn out to someday become persons? We should not kill people to benefit others, but we should also not let people die to protect human tissue such as sperm or ova, even though such gametes have great potential."[33]

THE PURSUIT OF PERFECTION

Conservative Christian thinkers and commentators, including a number of the president's advisors, brook no argument when it comes to morality and human forms of life. The message from members of this group is clear: "Leave the embryo alone." They believe that carelessly using an embryo for scientific research is the same as carelessly using a person for scientific research. Destroying an embryo is equal to killing a person.

It is not simply a case of Christian morality at work among those who object to hESC research. The connection to personhood recalls longstanding moral instructions from philosophers like Immanuel Kant (1724–1804), who said that persons should never be used a means for someone else's ends. The Kantian ethic of how we should treat each other is an essential ingredient of our professional codes of medicine—doctors must treat patients with dignity and respect and, above all, "do no harm." If an embryo is a person, then the Kantian imperatives apply.

Whichever side of the debate one chooses, dead philosophers help us only up to a point. It is no exaggeration to say that looking at early embryonic life in the glare of modern technology provokes a visceral response in some, and the feeling translates into a values-laden moral opinion. For example, Michael Gazzaniga uses his "gut feeling" and what he knows about science to determine that the embryo is not a person. The neuroscientist's political colleague was Leon Kass, former chairman of the PCBE. To Kass, the "gut feeling" is a sinking one. His reaction doesn't have much to do with whether the embryo has a moral status—Kass maintains he is "agnostic" on that point. A University of Chicago philosopher with degrees in medicine and biochemistry, Kass says our *hubris* is the danger—Americans have developed a scientific swagger that threatens our humanity. Because of our overconfident attitude, we find ourselves teetering on the precipice of a "slippery slope." Kass writes, "In leading laboratories, academic and industrial, new creators are confidently amassing their powers and quietly honing their skills, while on the street their evangelists are zealously prophesying a posthuman future."[34]

What lies at the bottom of Kass's stance that we have embarked on a perilous decent? An existence akin to Aldous Huxley's *Brave New World*, a futuristic place with an amoral society living in perfect health. *New World's* inhabitants have technology to blame for the lack of their humanity; children are born in embryo factories and tranquilizing drugs distributed by technocrats control adult behavior. Kass finds a deep resonance with Huxley's dystopic vision and compares it to our devotion to biomedical technologies, including those that deal with the human embryo. He begins with *in vitro* fertilization. The reproductive clinician, Kass believes, subverts family values by becoming a third partner along with the parents of the newly implanted embryo.[35] He is fearful that if left unchallenged, scientists will genetically engineer babies, grow "newborns" in laboratory vats, and use them as helpless factories to produce medical products and human parts.[36]

The embryos in his posthuman future are treated as "mere meat" and "human caviar."[37]

Kass admits his images are meant to provoke—Laurie Zoloth describes his views as a "philosophy that demands engagement." Though Kass avers that the embryo probably deserves less respect than a human being, he thinks we should treat it with dignity. After all, each of us begins life as an embryo and if *we* are worthy of dignity, then we are worthy of such respect at every stage of our existence. To experiment with an embryo is to meddle with nature, or worse, to play God. In this worldview, nature and God are intertwined and inseparable. Genetic engineering, genetic testing, *in vitro* fertilization, mood-enhancing drugs, and embryonic stem cells—indeed, any technology that pushes beyond "natural limits"—are listed as dangerous outcomes of our headlong pursuit of perfection.[38] In this very important respect, Kass is in agreement with conservative Christians about the embryo and hESC research. The trouble with biotechnology, writes fellow council member Michael Sandel, is that it represents "the one-sided triumph of willfulness over giftedness, of dominion over reverence, of molding over beholding." It leads, according to Sandel, a Harvard University professor of government, to a confusion "of our role with God's."[39] There is a distinction between using science for therapeutic reasons (antibiotics to overcome an infection) and using science for human enhancement (increasing our lifespans or our capacity for memory) concludes Meilaender. "Here is where the limitlessness of our desire enters in," he warns. "Enhancement deliberately tries to move beyond the normal condition of our species—and there is no end to the ways in which we might like to be 'better.' Either we get a grip on that desire or it will increase its already strong hold on us."

A cadre of commentators who worry about scientific "progress" echoes the *Brave New World* philosophy. If we aren't careful, they say, Huxley's fictional world may come to pass. Among the group is Daniel

Callahan, a co-founder of the Hastings Center for Bioethics, a think tank nestled in a tree-lined acreage just south of New York City. In his 2003 book, *What Price Better Health: Hazards of the Research Imperative*, Callahan writes that biomedical research has become a "moral imperative" with its promise to win the battle against the "diseases of aging." The spoils of war are longer, healthier lives. Our attempts to find ways of living longer shouldn't be a mandate. Callahan contends, "We ought to act in a beneficent way toward our fellow citizens, but there are many ways of doing that, and medical research can claim no more of us than many other worthy ways of spending our time and resources."[40] It's a fair argument: scarce resources can be better spent on preventing illness and disability and teaching people how to take care of their own health. Regenerative medicine seems to fall squarely within Callahan's crosshairs. He writes, "Unless someone can come up with a plausible case that the nation needs everyone to live much longer, and longer than the present steady gain of normalization will bring, there is no reason whatever for government-supported research aimed at maximizing or optimizing lifespans."

Joining Callahan and Kass is Francis Fukuyama. The professor of international political economy at Johns Hopkins University writes in the introduction to his book, *Our Posthuman Future: Consequences of the Biotechnology Revolution*, "The aim of this book is to argue that Huxley was right, that the most significant threat posed by contemporary biotechnology is the possibility that it will alter human nature and thereby move us into a 'posthuman' stage of history."[41] Another member of the president's council, Fukuyama warns, "New procedures and technologies emerging from research laboratories and hospitals ... can as easily be used to 'enhance' the species as to ease or ameliorate illness."[42] What are Fukuyama's remedies? In terms of the embryo, he sides with Kass: "I believe that human embryos ... are not the moral equivalents of infants, nor are they simply clumps of cells like any other tissue that can be discarded at will."[43] He argues for more regulation, but deems the American government unfit to oversee stem cell research. He cites how our present patchwork of regulations lag behind biomedical science's breakneck pace.[44]

TRUTH-TELLING

Among those who believe that hESC research should wait includes an ethicist who charges the real truth about stem cells has gone missing. The hype about cures—used by the media, stem cell supporters, and scientists alike—obscures the very real possibility that any cures of substance will be a long time coming, if they come at all. This is the sentiment of council member Rebecca Dresser. A professor at Washington University School of Law, Dresser maintains "truth-telling" is an overlooked dimension of stem cell ethics. "Scientists must be clear with patients, egg donors, and the public about the stage of this research. We don't know if it will produce treatments and cures," she says. "Research often surprises, and just as doctors should be honest about a poor prognosis or an uncertain outcome, researchers should be honest about all the things that remain to be looked at and tested."[45]

Dresser cites other, ballyhooed, scientific "revolutions" that have yet come to pass. "My skepticism comes from living through so many examples of excitement, including gene therapy and fetal tissue transplants. A lot of this stuff never works, and it takes a lot longer than people think. If it does produce an incremental benefit, it might only extend their lives for a few years." Dresser's statement might run afoul of critically ill patients, their families, and doctors who regard mere months of extra living as precious commodities. Yet looking deeper into the maw of healthcare reform and the best way to distribute scarce resources, her logic—along with Daniel Callahan's—demands a second thought. Why should we spend all this money at the leading edge of science? Wouldn't it be a much better use of resources to devise, for example, a fair and just means of distributing treatments we currently have to patients who need it? Dresser claims she isn't against hESC research per se. "Look, private funds can be used for stem cell research. And other countries permit it," Dresser states. "There are other exciting areas of research that could benefit from federal funding. It seems to me there ought to be a way to use the funding to benefit patients without violating the strongly held moral views of a large group of people in this country."

Along with nine other members of the president's council, Dresser voted in 2002 for a recommendation to place a four-year moratorium on embryonic stem cell research. At least in some council members' minds, the moratorium was designed to force scientists and supporters to write guidelines and rules to regulate hESC research. As the first anniversary of the PCBE recommendation nears, would Rebecca Dresser vote the same way again? She hesitates for a moment, and says, "Yes."

THE COSTS OF DELAY

"I don't believe for a second that voting repeatedly for a moratorium is any different than an outright ban."[46] This fiery comment comes from Alta Charo, an outspoken and energetic supporter of embryonic stem cell research. Like scientists who make it their business to know the experimental ins and outs of their competitors, Charo, a professor of law and bioethics at the University of Wisconsin, knows and studies the ethical positions of her peers, including the people she disagrees with. No wonder: as a discipline, bioethics is little bigger than even the youthful guild of stem cell biology.[47]

As some philosophers think "retreat"— that science should withdraw to a simpler, more comfortable time, Charo dares to think ahead. To her, the moral priority is the needs of patients, and she's leery of the government—or presidential advisors—slowing things down. She asks, "How do you balance the patient's demands against the government's role to provide safety?" Charo lists diseases where patient advocacy forced regulators to change (AIDS and cancer come to mind), and predicts the same will happen for stem cell research. "People with degenerative diseases will insist that clinical trials begin. Demand will come from parents with kids suffering from juvenile diabetes—nobody fights harder for a kid than a parent. There will be tremendous pressure on the Food and Drug Administration to quickly approve new treatments." She adds that patients and taxpayers alike will want returns on their political and financial investments in states that have passed legislation favoring stem cell research.

Charo's comments reflect the frustration of scientists who claim that politics, not ethics, stand behind pronouncements from the White House and the four-year-old executive order prohibiting government funding of new hESC lines. The criticism is that some government advisory bodies—including one with ethics as its focus—are anti-science. The 18-member PCBE originally had only three scientists: now there are two. Elizabeth Blackburn, a professor of biology and physiology at the University of California, San Francisco, was an outspoken opponent of the Bush administration during her rollercoaster tenure on the council. She was removed from the council in 2005. She firmly expresses her dissatisfaction with their deliberations in a paper entitled *Thoughts of a Former Council Member.*[48] "[A] moratorium is used to gain more information.... But that information can only be gained by performing the same research that the moratorium proposes to halt," she writes. Noting that adult stem cell research alone can't answer important questions about disease, she continues, "But one cannot find answers to questions about oranges by doing all of one's research on apples. Some research on apples will be useful because it will provide information that applies to fruit in general. Diseases, however, are very specific."

Blackburn, like Charo, regards putting science on ice while calling for more regulations as equivalent to maneuvering for an outright ban. "Such regulations may never emerge, allowing opponents ... to accomplish by administrative delay what they have been unable to accomplish by legislation, that is, a *de facto* ban on SCNT."[49] Along with Michael Gazzaniga, Janet Rowley—a geneticist and cancer specialist at the University of Chicago—is one of the two remaining career scientists on the council. She admitted in 2002 that stem cell treatments were largely based on promise. She asserts the reason that evidence was lacking that hESCs could treat Parkinson's and other problems was due to a decades-long government ban on funding embryo research.[50] Without embryos as a source of embryonic stem cells, knowledge is scarce.

Blackburn, Rowley, Charo, Zoloth, and other ethicists, theologians, and patient rights activists remind us that while the government dithers, people suffer. The reality of suffering is very much on Zoloth's mind. "I worked for 20 years as an intensive care unit nurse, and I know what it is to stand beside the body of a broken, sick child and be desperate for ways to heal the baby," she recounts. "Parents who inject their children over and over with insulin, constantly worried they will get it wrong; kids with sickle cell disease; people suffering from suicidal clinical depression—these are *not* trivial matters!" For Zoloth and other ethics scholars who call themselves "justice theorists," what matters is the priority given to certain kinds of medical research, and how best to distribute the discoveries among those who need it. Zoloth insists that medicine that cuts across class boundaries is the best medicine of all. The philosopher lists spinal cord injuries as one problem that plagues developed and undeveloped countries alike, and she applauds the use of stem cell lines to discover drugs for infectious diseases like malaria and tuberculosis, where pathogens make mincemeat of healthy cells.

Can an argument for longer lives with less suffering be reconciled with the contention that science has become a selfish means to a perfect end? Charo says, "Why would you say, 'all enhancements are bad?'" She uses a compelling object lesson. "If you could use genetic engineering to make your offspring resistant to a deadly virus, wouldn't you? Why wouldn't you?" The heated exchange between those who see biotechnology as out of control and those who don't has, according to Charo, stopped the discussion cold. "We need a nuanced, levelheaded approach about which alterations we are willing to accept," she says. The word "nuanced" however, doesn't occur to Arthur Caplan, chair of the Department of Medical Ethics at University of Pennsylvania, when he writes, "Beating up on the pursuit of perfection is silly."[51] We are already creatures of technology, says Caplan, and have been ever since we've been able to probe nature and question our relations to it. There is nothing in our history, he maintains, "that shows why we should not try to improve upon the biological design with which we are endowed. Augmenting breasts or prolonging erections may be vain and even a waste of scarce resources, but seeking to use our knowledge to enhance our vision, memory, learning skills, immunity, or metabolism is not obviously either."

Both Charo and Zoloth see the human responsibility to repair what nature has left unfinished, or as Charo puts it, "The point isn't that we play God, it's that we play human." To them, the goal is not to seek perfection. The task is—and always has been—an incremental, painful, and slow approach toward a better life. Zoloth says, "Thank God, we no longer die of smallpox, plague, and horrible infectious diseases. We should stop worrying about a tiny proportion of our society that might use new technology incorrectly. People are sick! Will some abuse science and medicine? Yes. They do already. I would rather let a few of those slip by than have hundreds or thousands of sick people not be treated."

GUIDANCE AND OVERSIGHT

Charo bristles at criticisms that hESC research needs to wait for better guidance—she's one of the principal authors of the National Academy of Science's (NAS) new 132-page report, *Guidelines for Human Embryonic Stem Cell Research*. "There is no way a moratorium can be lifted if the conditions for lifting it are that hESC research must have 'adequate regulations,'" she says. "All someone has to do is say, 'hESC research is inadequately regulated by government' to stop it. There is nothing easier for the government than *not* to act." The fact is, a government agency instructed *not* to fund new embryonic stem cell research—the National Institutes of Health—wasn't about to act anyway. So the NAS, a congressionally mandated group of expert scientists and engineers advising the government since 1863, did. The authors say the report is "designed specifically to provide comprehensive coverage of hESC research ... a system with both local and national components that meets the important goals identified by the other advisory bodies, including the President's Council on Bioethics in its report on somatic cell nuclear transfer."[52]

Among other things, the report recommends that each research center form an ESCRO (embryonic stem cell review oversight) committee, similar to committees that oversee and approve human clinical trial research. To garner a wide range of opinion, the NAS suggests an ESCRO should include a mixture of scientists, ethics, legal professionals, and members of the public. The committee's job is to ensure the research has scientific merit and that certain ethical guidelines are followed. If a scientist wants to make an embryonic stem cell line by using an embryo made by *in vitro* fertilization or by nuclear transfer, the ESCRO must be satisfied that the donors have given consent to use their cells or embryos for research and that no payments exchange hands between the donors and research centers. The group also is instructed to ask whether existing lines of hESCs can be used, recognizing that embryos shouldn't be destroyed unnecessarily.

Along with these questions, there are guidelines to protect the donors of cells and embryos. Donors should be assured that their confidentiality will be protected and told clearly that embryos will be destroyed during the process. A statement of risk is required, especially for women who donate eggs for research—a chemical procedure that stimulates the ovaries to produce extra eggs. Egg donation is an overlooked facet of stem cell ethics. Charo and her colleagues also suggest that both male and female donors be given the option of how they want their cells to be used. For example, some donors may feel fine about their embryo being used to make a cell line for research but uncomfortable if the cells are used to make organs or tissues for transplants.[53] If embryos or gametes come by way of infertility treatments, then the decisions to donate must be free of influence by any researcher interested in using them.

Committees that approve experiments on humans or animals, including hazardous procedures using radiation or infectious agents, have been part of research centers and teaching hospitals for decades. The knock on such groups from critics is that researchers are essentially policing themselves—a fox guarding a henhouse is little protection for the chickens inside. This is a conundrum for any "expert review" process, because often the best people to determine whether a project is scientifically, medically, and ethically sound are the experts themselves. Both Francis Fukuyama and Rebecca Dresser favor broadening oversight of hESC research to include greater participation by the public, although Fukuyama recognizes that members of society at large may not be knowledgeable enough about the science to participate meaningfully.[54] For her part, Dresser proposes something more unconventional. Rather than focus on protecting donors and determining the best medical need, she believes the first task should treat embryos as "moral scarce resources." "If we are serious about giving the embryo some respect, then we should have oversight committees that have an interest in treating them as precious," she states. "We call ourselves a deliberative democracy. Let's engage citizens who feel strongly about protecting an embryo. A real review should have teeth, where people who see the world differently can fight it out and argue about what's best." Both sides of the debate agree that citizen participation is necessary; the question remains what kind of public input will be most effective for ESCRO deliberations.

The NAS report contains other recommendations, including ways to create center-by-center stem cell registries and suggests that lines imported from other laboratories, especially those in other countries, are obtained in ways consistent with the ethical guidelines set down by the NAS. Some of the report follows Britain's lead, where embryo politics aren't so fractious. In the United Kingdom, a centralized agency, the Human Fertilisation and Embryology Authority (HFEA) regulates the research by licensing hESC lines to investigators. The NAS suggests a national authority like the HFEA be organized in the United States. Like the HFEA, the NAS guidelines prohibit researchers from keeping embryos alive past fourteen days, when the neural crest appears.

CAN MICE DO CALCULUS?

Time will tell whether the NAS report will take hold. For now, it represents the only attempt at a national set of hESC guidelines. What surprised many was the report contained a section on a controversial dimension of stem cell research: animal-human chimeras. Different than the human-human blood chimeras discussed in Chapter 7, this chimera results from human stem cells transplanted into a lab animal. Animals and humans already share cells and tissues. Heart valves from pigs replace human valves damaged by heart disease. We can do experiments on specially engineered strains of mice with a human immune system that would not be permitted on humans. The ethical equations change, however, when transplanting human embryonic or neural stem cells into animals. If a mouse with a functioning brain made entirely of human neurons can be produced, it could be used as an experimental model for Parkinson's, Alzheimer's, or other dementias. New drugs or other treatments could be first tested for their effects on mice before moving on to human clinical trials. How human neurons are damaged by pathogens could be studied in this mouse without risking harm to a human subject.

Trouble is, a mouse with a functioning brain made from human neural stem cells may no longer be a mouse. It could be partly human.

Such a creature would be vastly more mouse than human, experts contend. A mouse's brain is 1/1,000th the size of a human brain and, as a consequence, organized much differently. Neuroscientists claim the differences in "architecture" among animals as the principal reason that mice are not men—or women.[55] Without proof otherwise, experimental animals made with human stem cells could have an altered form of consciousness. This area of neurosciences is still in its infancy, so experiments to make a neural-mouse model will no doubt be closely monitored by ethicists and neuroscientists who wonder what degree of brain complexity is needed to trigger consciousness or even sentience, the ability to feel pleasure and pain.

The NAS guidelines don't prohibit putting human stem cells into an animal brain; but they do stipulate the degree of caution rises as scientists move up the evolutionary ladder. For example, work with primate embryos raises red flags. Putting a human embryonic stem cell into a monkey blastocyst is forbidden because it will integrate and contribute to the development of the primate—resulting in a human-primate chimera. Such an animal may have heightened consciousness, appear human, or have human characteristics embedded in the genes of eggs and sperm, outcomes unnerving to imagine and of deep concern for most people. Certain human-to-human neural stem cell transplants could be quite controversial; fussing with the human brain could positively or negatively affect the way we process sensory information. "We don't know whether we are on thick or thin ice with neural stem cell transplants," Charo warns.

POLICY OR JUST POLITICS?

Ethics isn't limited to street corners, coffee shops, and university corridors. In its most mature form, ethics becomes policy—rules made by society to guide our best attempts to living a good life. More than ever before, positions pounded out by ethicists are being used for political ends, especially in government administrations. The debate has reached every developed country with the capability of biomedical research. What are the consequences of limiting or outlawing embryonic stem cell research? How does the U.S. compare do other governments facing the same challenges?

The moral divide between philosophers has widened to a chasm between politicians. Our state and federal governments have changed—indeed are now changing—our ability to use embryonic stem cells. In consecutive congressional sessions, legislators considered a law to criminalize embryonic stem cell research. States such as California and New Jersey have passed their own laws permitting stem cell research. In the next chapter, we explore how well our public ethics fares when it intersects politics and law.

9

The Consequences of Politics

> In the end, politics will settle the debate in this country about whether human therapeutic cloning is allowed to proceed. If the decision is yes, then we will continue to lead the world in a crucial, cutting edge area of biomedical research. If it is no, U.S. biologists will undertake hegiras to laboratories in Australia, Japan, Israel, and certain countries in Europe—an outcome that will leave American science greatly diminished.[1]
>
> NATIONAL MEDAL OF SCIENCE WINNER ROBERT WEINBERG

James Thomson's announcement of the successful isolation and proliferation of human embryonic stem cells in November 1998 blew the lid off a long-simmering controversy about embryo research and whether rogue scientists would try to create human clones. The 1973 Supreme Court ruling in *Roe v. Wade* legalizing abortion in the first two trimesters of pregnancy opened a Pandora's box of worst-case scenarios. Religious leaders and a vigorous anti-abortion movement claimed the decision would result in the black-market sale or barter of fertilized eggs and indiscriminate use of embryos for laboratory experiments. Even more worrisome to the pro-life supporters was the possibility of the unregulated use of aborted fetal tissue. An alarmed Congress halted federally funded embryo research until guidelines could be established.[2] The 1974 action had surprising staying power. With one short-lived exception, the "temporary" moratorium has passed its thirtieth anniversary—no government funds are allowed for embryo research, a policy that swept essential questions about infertility, reproductive medicine, prenatal diagnosis, and embryonic stem cell research beyond the reach of most American clinicians and scientists.

With U.S. reproductive research in arrears, it was technology from England that offered the first glimmer of hope to parents who could not otherwise conceive children. After nine years of trying to become pregnant, Mrs. Leslie Brown gave birth to a baby girl, Louise, on July 25, 1978. The Browns had enlisted the help of R. G. Edwards, a Cambridge physiologist and Patrick Steptoe, an obstetrician in private practice. Edwards and Steptoe had perfected a procedure for retrieving oocytes, fertilizing them in the laboratory, and then implanting the embryo back into the womb. Upon her birth, Louise was hailed as the world's first "test-tube baby." The Vatican promptly condemned the procedure and one of the first voices attempting to elaborate American ethical policy on *in vitro* fertilization (IVF) came from Leon Kass, who in 1979 published his opposition to assisted reproductive technologies in several articles.[3]

In the coming years, human embryo research languished in political purgatory. The Department of Health, Education, and Welfare (now Health and Human Services, or HHS) and its ethics advisory board attempted to draft policy within the context of IVF, and in 1979 determined that "the human embryo is entitled to profound respect, but this respect does not necessarily encompass the full legal and moral rights attributed to humans" and further recommended that its subordinate organization, the National Institutes of Health, fund research on extracorporeal embryos up to 14 days of age.[4] These recommendations were summarily rejected by the HHS and never became law. Seven years on, encouraged by animal experiments using fetal tissue, one of the NIH's own team of investigators applied for funds to transplant fetal cells from elective abortions into Parkinson's patients. Once again, the HHS rejected the request and in 1987 banned fetal tissue research outright. This time, Sweden stepped into the breach. Anders Bjorklund (Chapter 7) and his group at the University of Lund successfully transplanted fetal cells into Parkinson's patients, marking the first therapeutic use of cell therapy for brain disease.

Bjorklund's results and prevailing winds from laboratories outside America prompted the NIH to form a panel to review the ethics of fetal tissue research. In 1988, it voted 18 to 3 in favor of funding both embryo and fetal research, arguing that use of fetal tissue to treat disease is distinct from the morality of abortion. Despite the ruling, HHS Secretary Louis Sullivan—a George H. W. Bush appointee—extended the moratorium indefinitely, bowing to the fears of three dissenting religious conservative panel members that such research would increase abortions.[5]

Ethics, science, and religion became further intertwined with politics in the 1990s, increasing the battles among Congress, administrations, and their agencies. In 1990, Congress tried to override the 1974 ban, only to have the senior Bush veto the action. Spurred by the efforts of patients' rights and disease advocacy organizations, the window in support of embryo research opened briefly in 1993. President Clinton issued an executive order instructing HHS secretary Donna Shalala to lift the congressional ban. Both an internal NIH committee, the Human Embryo Research Panel, and Clinton's own ethics advisors recommended that research begin using donated IVF embryos and aborted fetal tissue. The panel stated, "The promise of human benefit from [embryo] research is significant, carrying great potential benefit to infertile couples, families with genetic conditions, and individuals and families in need of effective therapies for a variety of diseases."[6]

Clinton quickly backtracked after he became concerned over thousands of letters sent from pro-life supporters. The NIH didn't proceed with funding because it feared congressional backlash. The presidential and agency waffling was all Congress needed. In 1995, it put teeth to the ban in the form of a funding rider renewed every year since. The Appropriation Act states that taxpayer money may not be used for the creation of human embryos for research purposes or research in which embryos are destroyed. James Thomson and other American reproductive and developmental biologists had to eke out funding for embryonic stem cell discoveries using private dollars from philanthropy or corporations.

CHANGING ADMINISTRATIONS; CHANGING WINDS

During the waning years of the Clinton administration, Shalala's HHS determined that though the ban stated that no embryos could be destroyed with government dollars, NIH funding could be used for research on hESC lines established with private dollars. This meant that if money from a philanthropist or corporation was used to make an hESC line, it was permissible to use additional government resources to, for example, use the same line to make neural stem cells. The NIH followed with a set of draft guidelines in December 1999. Research would be permitted on embryos that remained after fertility treatments, provided they were given with the informed consent of the donors and the fertility clinics made no profit from the exchange. The National Bioethics Advisory Commission (NBAC), created by an executive order from Clinton, weighed in and went one step further: they provided a list of reasons why the ban should be lifted, including shortening the time to clinical trials and promoting competition among biotechnology companies in order to drive down health costs. With respect to the moral status, the NBAC wrote, "The embryo merits respect as a form of human life, but not on the same level accorded to humans."[7] The commission warned that in contrast to the NIH, private sector companies had no obligation to make hESC research available to the public.[8] Keeping embryo research in the public domain would ensure transparency and equal access among many laboratories. Final guidelines were issued and approved by the president on August 25, 2000, and the NIH began soliciting applications for research grants. Some proposals came in, but not nearly as many as had been anticipated, partly because of remarks made by then-Governor Bush during his 2000 campaign suggesting that, if elected president, he would promptly reverse Clinton's policy.[9]

In fact, President Bush did not immediately reverse the policy; shortly after taking office in 2001, he called for yet another HHS review. The NIH abandoned its plans to review grant applications—proving the wisdom of scientists who had not wasted their time on dubious hopes—and the debates began anew, with scientific organizations, companies, patient advocacy groups, and religious organizations slugging it out on widely divergent and constantly shifting moral, scientific, economic, and medical grounds. Industry lobbying organizations, like the drug industry's Pharmaceutical Research and Manufacturers of America and the Biotechnology Industry Organization were, for once, curiously restrained, apparently sensing that discretion was the better part of valor.

In a televised address on August 9, 2001, President Bush presented his decision. Federal funding could be used to research only cell lines created before his televised address at 9:00 P.M. In addition, the embryos from which the stem cells come must have been created for reproductive purposes, must no longer be needed for reproductive purposes, and must have been obtained with informed consent and without financial inducement. Months later, the NIH released a list of the 10 worldwide organizations that were holding the 64 cell lines—later revised upward to 78—that met the president's criteria and, thus, were eligible for federal funding. On the heels of that proclamation, he appointed Leon Kass as head of the President's Council on Biomedical Ethics (PCBE) to further advise him on this and other ethical issues. Eighteen advisors were appointed to the council, most of them academics with expertise in law, theology, political science, economics, and traditional medicine. The choices caused worry among basic scientists and ethics professionals. University of Pennsylvania's Arthur Caplan, considered one of the deans of American bioethics, criticized the makeup of the council, saying, "They've got some intellectual heavy-hitters there from the right wing, but not people who are necessarily conversant with current developments in biomedicine."[10]

Later that year, the House of Representatives followed the White House lead and, by a vote of 265 to 162, chose to ban the cloning of humans and criminalize nuclear transfer. Under a bill (H.R.1357) sponsored by Representative Dave Weldon (R-FL), any scientist caught using the technique would be subject to a penalty of $1 million and up to ten years in jail. After the House passed the legislation, it was given to the Senate for consideration. In January 2002, Senators Sam Brownback (R-KS) and Mary Landrieu (D- LA) introduced a proposal in support of the House's bill. The Brownback and Weldon bills contained a particularly onerous provision, mandating the same criminal penalties on any American who *provides or receives* medical treatments involving SCNT technology developed in another country. It also meant that patients and doctors could go to jail, too. These actions met with nearly universal opposition in the scientific community. Eighty Nobel Prize winners, led by Stanford's Paul Berg, wrote a letter to the president condemning the policy, arguing that it was crippling their research. Berg, a man who has spent a career at the forefront of science policy, understands how important it is to measure words. But his decorum disappears when he describes how he felt when he first read the Weldon legislation. "I couldn't believe the arrogance of a bunch of people in Congress saying to 290 million Americans, sorry folks, you're not going to have the therapies to cure your disease because we are offended by this technology."[11]

In an April 2002 press conference, Bush (who had previously described therapeutic cloning as "growing human beings for spare body parts") said, "I strongly support a comprehensive law against all human cloning. And I endorse the bill—wholeheartedly endorse the bill (S. 658)—sponsored by Senator Brownback and Senator Mary Landrieu. This carefully drafted bill would ban all human cloning in the United States, including the cloning of embryos for research."[12] When the Senate failed to vote on it and left the bill for the 108th Congress to reconsider, Senator Brownback switched to a backup plan: he maneuvered the bill out of the purview of the Judiciary Committee, whose pro-life chairman, Senator Orrin Hatch (R-UT), supports stem cell research. Both Hatch and Arlen Specter (R-PN) broke ranks with conservative Republicans, and in a press conference Hatch announced his support of the competing bill (S.1602), sponsored by Dianne Feinstein (D-CA) and Edward Kennedy (D-MA), that prohibited reproductive cloning but allowed cloning of embryonic stem cells. Hatch wrote in a press release, "I come to this issue with a strong pro-life, pro-family record. But I also strongly believe that a critical part of being pro-life is to support measures that help the living."[13]

The ensuing political scrum produced a flurry of competing legislation. In 2003, the House introduced five stem cell bills, the Senate, two. During the debate and testimony, every major American scientific body weighed in. The National Academy of Sciences, the American Association for the Advancement of Science, the American Medical Association, patient advocacy groups, and a growing list of legislators lined up to protest the Administration's policy, including 11 house Republicans who wrote to express their concerns about the quality of cell lines on the NIH registry. In May 2004, 206 members of the House—including three dozen opponents of abortion and many conservative Republican leaders—wrote a letter to Bush urging him to loosen the restrictions. In June, 58 senators followed suit with a similar plea. Senator Bill Frist (R-TN), an original supporter of the president's policy, said he wanted it reviewed. He said, "I'm very interested in answering the question whether or not scientists are really leaving the country in droves because of the limitations on research."[14]

Bush remained undeterred. An HHS press release issued late that summer, titled "Stem Cell Facts," made it clear his position was a simple matter of conviction and principle. With emphasis it states, "The president is committed to pursuing stem cell research without crossing a *fundamental moral line* by providing taxpayer funding that would sanction or encourage further destruction of human embryos."[15]

In the last half of 2004, a year that featured a presidential election, the political din was deafening. In June, Ronald Reagan lost his long battle with Alzheimer's disease, reminding everyone how his wife Nancy had two years earlier quietly but firmly told the White House of her frustration with the NIH stem cell policy. More than that, she had used her old Washington network of friends and associates to lobby Congress on behalf of Hatch and Specter's Senate effort to authorize therapeutic uses of hESCs. Then Ron Reagan Jr. surprised everyone by speaking about the promise of stem cells at the Democratic National Convention in late July. The speech was impassioned and accessible. Furthermore, the pundits who predicted that stem cell research would be nothing more than a back-burner election issue were proved wrong. After Ron Jr.'s speech, both Bush and democratic challenger John Kerry placed the issue in their quiver of campaign messages.

At the height of the campaign, Christopher Reeve died—his body had given up under the stress of severe quadriplegia caused by an equestrian accident ten years earlier. His tireless advocacy for stem cell research from the confines of his wheelchair demonstrated he still had the same indestructible quality as his movie characters: the tenacity and refusal to acknowledge anything less than total commitment to cures for spinal cord injury. It had forged him into an archetype for patients and advocates frustrated by science's slow progress and the impediments of politics.

Candidate Kerry used both events to promote his science policy. Kerry, a Catholic, said if elected he would overturn the congressional ban and increase the funding for hESC research: "Patients and their families should no longer be denied the hope that this new research brings."[16] For Bush, stem cells and embryos were part of a larger agenda, tightly wound to the moral certainty of his religious faith, family values, and opposition to gay marriage and abortion. Bush knew how important it was to define a political-moral framework that would appeal to conservative Christians and evangelicals and in the end, morality proved to be a deciding gambit. In every category of religion except Jewish voters, Bush captured strong majorities, increasing his percentages from four years earlier. Exit poll respondents called "moral values" the election's most important issue, more important than the economy, terrorism, or the war in Iraq.

After the election, the second session of Congress was deadlocked on the issue and did not pass the bill in 2004. In early 2005, a quartet of stem cell legislation lay before the 109th House and Senate. All expressly banned reproductive cloning. Only Weldon's bill (H.R. 1357) had passed the House, and both it and Brownback's bill not only ban somatic cell nuclear transfer (SCNT) but criminalize an entire branch of scientific study. On the pro stem-cell side, the Feinstein-Kennedy bill included Republican co-sponsors Orrin Hatch and Arlen Specter and permitted the use of SCNT to derive human embryonic stem cells. Republican John Danforth, a former senator from Missouri and an Episcopal minister who resigned in January 2005 as ambassador to the United Nations weighed in with an essay in *The New York Times*. He warned, "Republicans have transformed our party into the political arm of conservative Christians," and "[T]he only explanation for legislators comparing cells in a petri dish to babies in the womb is the extension of religious doctrine into statutory law."[17]

In the spring 2005, the conservative-minded House began to swing slowly in favor of hESC research. Representative Michael N. Castle (R-DE) introduced H.R. 810, the Stem Cell Research Enhancement Act of 2005. The bill would require NIH to fund research on new stem cell lines derived from embryos discarded by fertility clinics, essentially overriding the president's 2001 policy. In late May, the house passed the legislation, 238 to 194—not enough to avoid a presidential veto. Soon after, President Bush threatened to veto any legislation permitting hESC research. After receiving the bill from the House backers, Arlen Specter said, "I don't like veto threats, and I don't like statements about overriding veto threats." Speaking at a news conference assembled for the occasion, he went on to say, "I think if it really comes down to a showdown, we will have enough in the United States Senate to override a veto."[18] In July 2005, the Senate Republican leader Bill Frist finally made a clean break with presidential politics, throwing his support behind the Stem Cell Research Enhancement Act. He declared before Congress that stem cell research "isn't just a matter of faith, it's a fact of science."[19] By mid-2006, stem cell and cloning legislation had multiplied like stem cells themselves—16 bills waited for a vote. Frist promised to schedule the Senate votes during the summer months.

THE HIGH COUNSELORS

Appointed commissions that advise presidents on bioethics and science policy are nothing new. Arguably, the most productive group was Richard Nixon's National Commission for the Protection of Human Subjects of Biomedical and Behavioral Research. Made up of lawyers, physicians, ethicists, and scientists who made a point of engaging regular citizens in its deliberations, the commission vigorously produced policy and guidelines that have since become standards for the ethical conduct of science. Besides creating the towering Belmont Report in

1978, which fast became the law that protects patients who participate in clinical research, the commission published tracts on research involving fetuses, children, and prisoners, as well as guidelines for the delivery of health services. Albert Jonsen, an ethicist on the National Commission, recalls the high road it took during its deliberations: "The reports resulted from a group of people—some professionals, some laypeople—working on an ethical problem and arguing about it in a sustained, public way. This practical approach to ethics was its finest achievement."[20]

The success of the first commission prompted the Senate in 1978 to renew and expand its mandate. In an important legislative twist, it required that its members be presidentially appointed and that the Senate confirm the chairperson. This structure has been a distinguishing feature of ethics councils from the Carter presidency onward. It allows chief executives to populate their commissions any way they wish, and though résumés of Bush's Presidential Council on Bioethics (PCBE) reflect a diversity of backgrounds, the appointments reflect an aggressive use of that prerogative. Since 2002, three members have left the council—evidence of further political shuffling. One resigned, another's term was not renewed, and a third was removed. All three supported embryonic stem cell research. Kass stepped down as chairman in the Fall of 2005, but remains on the council.

It is not clear whether history will paint the PCBE with the same respect it lavished on the National Commission of the 1970s. The council did get off to a rough start. Accusations abounded that the PCBE used political litmus tests to screen prospective commissioners, gerrymandered votes, and cloistered sessions without public input. To be fair, the council's early opinions reveal a healthy degree of debate, especially after the PCBE published its first recommendations on stem cell research in July 2002 in a report called *Human Cloning and Human Dignity*. Seventeen members (with one abstention) voted to ban "cloning-to-produce-children." Whether to recommend "cloning-for-biomedical-research" (including nuclear transfer) was a different matter entirely: the council faced a near deadlock. The faxed votes began to roll in: seven members voted for a ban and seven members voted to permit hESC research. Three members favored a wait-and-see approach, recommending a four-year moratorium. The personal statements made by members reveal that significant moral wrangling went on behind the scenes. The council had to break the deadlock and did so by reaching a compromise. The final tally: ten votes in favor of a four-year moratorium; seven votes against.[21]

After the recommendations were issued, *Science* published two essays, one from a group of top-ranked scientists and the other from its editor-in-chief, Donald Kennedy. Both pointed to the singular tendency of the Bush administration to stack its science advisory committees with people who share its views, contrary to federal law, which requires advisors to represent balanced perspectives. Testimonials from PCBE member Elizabeth Blackburn and other short-listed committee nominees bolster these claims, and they recount being grilled by White House officials about who they voted for in the 2000 election, their religious leanings, and whether they held views that "might be embarrassing to the president."[22] Such litmus tests are not limited to the PCBE. The Union of Concerned Scientists and Representative Henry Waxman (D-CA) released separate reports detailing political interference in a number of science agency appointments, including the NIH, NASA, EPA, and the FDA.[23, 24] Even the World Health Organization complained it had to raise its hand for HHS permission every time it needed a U.S. government advisor, rather than contacting experts directly as it has in years past.[25] *The New York Times* reported how career scientists and officials within the government had begun to complain about the heavy-handed tactics.[26] Alden Meyer, the director of the Union of Concerned Scientists, declared at a National Academy of Sciences meeting, "It's clear that if this pattern of abuse continues, there will be an exodus of scientific talent from the federal government."[27]

Despite the swirling controversy, the council has pushed ahead. The group has produced four major reports and solicited input from experts with opposing views. Productivity aside, the lingering criticism of the PCBE is that it too aggressively pursues a predefined, religious conservative agenda. The grumbling started early when conservative council members, including Kass himself, broadcast their ethical positions on TV and radio prior to the first meetings, saying that they reflected personal, not official views. To some, the media spots were clear evidence of a conflict of interest. Since then, the moral overtones of the reports are consistent with longstanding opinions of the majority council members, reflecting the "Brave New World" philosophy: left unchecked, big science, stem cell research, and biotechnology threaten the humanity of Americans.

This predeterminism and lack of balance has caused the council minority to dissent loudly and often. In September 2002, councilors Blackburn, Rowley, Foster, and Gazzaniga called the first report "short sighted" and urged Congress to lift the ban and "remove impediments to this critical research."[28] University of California, San Francisco, professor Elizabeth Blackburn, an award-winning and internationally known cancer expert, voiced her deep disapproval of the 2004 PCBE *Monitoring Stem Cell Research* report. In an April essay in *The New England Journal of Medicine*, she revealed how the dissenting opinion she submitted was not included in the council's publications. In another paper, Blackburn and council member Janet Rowley, a University of Chicago geneticist, claimed the content in the report lacked neutrality and "distorts the potential of biomedical research," including incomplete or incorrect descriptions of stem cell plasticity and the promise of adult stem cells.[29]

The president's response to the discordant notes was to dump Blackburn. He also asked William May, a Southern Methodist University ethicist and critic of the administration's positions, to wind up his term on the council. That left Rowley, a Lasker Award winner—an honor considered a bellwether to the Nobel Prize—shouting into the wind. She did so with gusto in July 2004: "I have seen first hand through the president's council that this administration distorts scientific knowledge on stem cell research, which makes it increasingly difficult to have an honest debate in a field that holds promise for treatment of many serious diseases like Parkinson's and juvenile diabetes."[30] The reply from the president's spokesperson was simply that Bush had "decided to appoint other individuals with different expertise and experience."[31] James Wilson, who in 2002 voted for embryonic stem cell research resigned nine months later. The three replacements—two political scientists and a pediatric neurosurgeon—are considered critical of stem cell research. Later Rowley said in an interview with the *Washington Post,* "I think this is Bush stacking the council with the compliant."[32]

DISAPPOINTMENT IN THE NIH REGISTRY

In an apparent concession to the promise of embryonic stem cells, Bush's restriction does not apply to any line made before his fateful August 9, 2001, proclamation. Those cell cultures presumably created between Thomson's 1998 paper and the 2001 address are now listed on the NIH's Human Embryonic Stem Cell Registry. Embryonic lines can still be made using private dollars, but even these "legal lines" are barred from appearing on the registry and from receiving federal support for their study. Thus for the vast majority of government researchers with little or no access to alternative funds, the official list is all that's available. In a cruel coda to the funding ban, it has turned out to be a vanishingly thin resource. NIH officials say only 23 of 78 lines will ever be available for research. Of the remaining 55 unusable lines, 16 perished after being removed from the freezer. Seven of the lines are duplicates, one is still in development, and one was withdrawn.[33] Thirty-one lines exist in institutions outside the United States, and according to James Battey, the NIH official who initially administered the registry, "We have no indication that any of these institutions will ever seek NIH support to develop their lines, or will make any effort to distribute their lines to the research community."[34]

How did the nation's top biomedical research agency, accustomed to providing billions of dollars of support and scores of research tools to government-sponsored scientists, manage to mishandle such a high-profile resource? When asked how the administration arrived at the original number, Battey said, "I have no idea."[35] After Bush's announcement, there was a scramble to come up with a list of stem cells to offer to scientists. It took several months for the registry to go public, and when it finally did, it bore an uncanny resemblance to a list published months earlier by *The New York Times*. Nobel Laureate Paul Berg, who has been at the forefront of policy battles on behalf of stem cell proponents, recalls his frustration when he finally saw the registry. "The only difference between the *NYT* list and the registry was that the NIH lines had a number assigned to them—no data, no other information at all. I called them and said, 'You can't be serious! This is the registry you've been bragging about?'"[36] According to Berg, NIH administrators told him that getting approvals from the White House had further delayed the release. Hardball politics also played a role. Republican Senator Arlen Specter—who was later diagnosed with Hodgkin's lymphoma—had introduced an amendment that would cause the Bush prohibition to disappear if the president was voted out of office. The White House, offended by the move, threatened to hold up the registry until the amendment was removed. Specter eventually withdrew the amendment, and the list went public.

The sobering reality is that many of the original lines can't be located, are of spotty quality, or are hopelessly tied up with patent rights. What's more, none of the lines are suitable for medical purposes. Fred Gage of the Salk Institute in La Jolla, California, and Ajit Varki of the University of California, San Diego, examined all the available cultures. Unlike colonies grown using recent techniques, they found the Bush-approved cells were cultivated with mouse feeder layers.[37] The fear is that mouse viruses contaminating the lines could also infect humans if they were used for transplants. The pair also found that human antibodies aggressively attack hESCs grown with mouse cells, which could lead to more severe immune rejection after the cells are transplanted. To make matters worse, the lines also have HLA profiles that make them unsuitable for widespread use as therapeutic transplants. In another study completed at the University of Washington, at least five of the lines were found to be nearly worthless because they are so difficult to grow.[38] Finally, the lines are of limited genetic diversity. Those couples donating embryos through IVF clinics were largely upper-class, white, and infertile. Many diseases such as Tay-Sachs, sickle cell anemia, and type II diabetes are found in specific ethnic populations, and developing lines for these disorders requires using embryonic cells taken from a matched genetic pool. During congressional testimony in April 2005, NIH agency heads criticized the registry as inadequate and warned that the United States was rapidly losing ground to other nations. "Progress has been delayed by the limited number of cell lines," wrote Elizabeth G. Nabel, the new director of the National Heart, Lung, and Blood Institute. "The NIH has ceded leadership in this field."[39]

THE WORLD STAGE

The United States is accustomed to leading the world in cell and molecular biology. But a survey of the most recent international stem cell meetings show impressive scientific results accumulating from geographies where hESC work is allowed: Israel, the Pacific Rim, Northern Europe, and the United Kingdom.[40] Of the currently available lines on the NIH registry, most were developed on foreign shores, a stark reminder of the long-term effects of American policy.[41] Countries with growing research efforts, including Singapore, the Czech Republic, Russia, Iran, Sweden, Finland, South Korea, and Israel, have begun to announce the availability of new lines. Some lines have been derived in conditions free of animal cells. New hESC lines are obviously not on the NIH registry and the rate at which they are announced is gaining momentum—of the 100 or so non-NIH lines available for research, most were added in 2004 and 2005 and more than half came from outside the United States.[42]

Of the countries that permit the use of embryos, many have passed laws or use guidelines regulating their research. In 2001, a bioethics commission in Israel approved the use of stem cells from early embryos and aborted fetal tissue up to nine weeks of age. The country followed the legislation in 2004 by forming a multimillion-dollar public-private consortium uniting five stem cell companies with the country's leading academic laboratories. All seven parties of the Swedish Parliament reached consensus on whether to permit the use of stem cells. Considered a stem cell powerhouse, Sweden boasts over 300 stem cell scientists working at nine centers, including Göteborg University and the highly regarded Karolinska Institute. Singapore's Bioethics Advisory Council issued a five-part recommendation that now guides the country's approach to stem cell research. It requires hESCs to be taken first from existing lines and if necessary, from embryos less than 14 days old. It allows the creation of embryos for research purposes on the basis of strong scientific and medical merit, and only after a statutory review.[43] However, a high-profile stem cell consortium made of European Union members has stalled because of the differences between nations with liberal stem cell regulations can't reach agreement with those nations with more conservative nations such as Germany and Italy.[44]

The Mexican parliament recently formed the National Institute of Genomic Medicine, which will use SCNT and human embryonic cells for research purposes. In March 2005, Brazilian President Luiz Inacio Lula da Silva signed legislation permitting research on over 20,000 frozen embryos from fertility clinics.[45] Australia's 2003 cloning act allows scientists to develop hESC lines from extra IVF embryos but only under government license. It criminalizes human cloning and prohibits Australian scientists from using nuclear transfer to create embryos for research. In July 2004, the French parliament passed legislation that allows SCNT but bans cloning humans. Canada still struggles to define its cloning legislation, but permits hESC research under certain conditions.

Britain was the first to authorize nuclear transfer and it now licenses SCNT technology developed by the government to scientists. In the United Kingdom, the National Institute for Biological Standards and Control opened a stem cell bank in 2002 to house human embryonic and adult stem cell lines from any source and provide them to any company or institution who demonstrates evidence of ethics committee approval, peer review, and informed consent for individuals taking part in any clinical trial involving the use of a human stem cell line. The bank has earned high marks both from scientists and from forward-thinking governments looking for a best practice for a national registry. The well-characterized lines are kept in high-quality conditions and are quality controlled for both "research grade" and "clinical grade" uses. The bank has more than 30 lines available, including lines from the Stem Cell Biology Laboratory at the Centre for Neuroscience Research, King's College London. In a 2004 interview, the center's director, Stephen Minger, said of U.S. restrictions, "It is the most intellectually incoherent policy you could possibly have. Britain is going to lead the world in stem cell research. I'm absolutely convinced of that."[46] In one crucial measure of scientific progress, it isn't Britain leading the way, it is the Pacific Rim. The percentage of hESC research papers coming from Asia jumped from zero to more han 20 percent in the period 1999–2004, while research from Europe and America declined. As a category, hESC research lagged other American biotechnologies by a significant margin.[47]

No doubt the UK's Minger expresses similar irritation with Germany, which forbids the production of new hESC lines and limits research to lines made before 2002. It does, however, allow importation of lines into Germany made elsewhere. In expected fashion the United Nations can't agree whether to ban or allow nuclear transfer. In a protracted, three-year discussion, all 191 countries agreed that human clones are a bad idea. In 2003, a vote to also ban SCNT was postponed for more deliberation after a dozen leading stem cell scientists gave a scientific symposium to the voting delegates. Since then, about 30 percent of the voting countries—mostly Roman Catholic and developing countries led by Costa Rica—continue to favor a total ban. When the issue came up for deliberation again in October 2004, the White House scuppered a consensus drive favoring nuclear transfer led by Belgium and Great Britain. Bush Special Advisor Susan Moore averred that scientific progress was possible without posing a "threat to human dignity" and allowing SCNT would "authorize the creation of an human embryo for the purposes of destroying it, thus elevating the value of research and experimentation above that of a human life."[48] In March 2005, the U.N. finally voted to ban "human cloning." The nonbinding declaration was widely criticized by both stem cell supporters and opponents as toothless and muddled.

GOING IT ALONE

For American scientists, a pandemic of politics was enough to cause some eminent researchers to abandon their laboratories. But the restrictions caused one stem cell expert with a steely resolve to find a way around the barriers. There is more than a career at stake for Harvard's Douglas Melton: his son and daughter suffer from juvenile diabetes. "I have two roles: one is as a scientist working on a cure for diabetes, but then like millions of parents, I unfortunately have two children who suffer from the disease. Like any parent, I will do anything I can to try to find a cure," he said in a CBS news interview.[49] Melton and his research partner, Yuval Dor, use adult and embryonic cells to investigate autoimmune disease. Their experiments (featured in Chapter 6) illustrate how important it is for both adult and embryonic stem cell research to proceed. If the pancreas doesn't have a resident stem cell, as the duo's experiments apparently show, then an hESC line will have to serve as the source for insulin-producing cells. To hedge his bets against this possibility, Melton began a joint project in 2002 with a Boston-based infertility clinic that supplied 344 donated embryos left over from fertilization procedures. Then he asked the Howard Hughes Medical Institute (HHMI) to build and equip a separate laboratory in which to do the research. HHMI, Harvard, and the Juvenile Diabetes Research Foundation anted up the millions necessary to conduct the science. In March 2004, he announced he had made 17 new hESC lines, nearly doubling the number of usable lines on the NIH registry.[50] And in the spirit of open access, he gives them free of charge to any researcher who requests them. A year later, he had shipped hundreds of cultures to researchers around the world.

Like the multimillion-dollar stem cell programs launched earlier at Stanford University and the University of California, San Francisco, Harvard took the opportunity to announce a month later the formation of a new enterprise, the Harvard Stem Cell Institute (HSCI). Melton and blood stem cell clinician David Scadden are leading a group of 30 faculty distributed across 7 Harvard schools and 8 affiliated hospitals. Their goal: repair organs riddled with disease, with emphasis on the pancreas, bone marrow, nerves, and the heart. Recently, Melton has ventured into controversial territory—the production of embryos specifically for research. Melton was the first to ask to produce embryos using SCNT that will give him better experimental control. After a year of ethical review, Harvard gave him the go-ahead to make embryos using nuclear transfer. He plans to develop disease-based lines using the nucleus of somatic cells from patients with type I diabetes, Parkinson's disease, and Alzheimer's disease. The resulting disease-in-a-dish promises to become a powerful tool for studying genetic illness and for discovering new drugs.[51]

Unfortunately, 10, 20, or even 50 million dollars here and there from institutions like Harvard, Stanford, and Columbia (which announced in June 2005 it was halfway to a $50 million target for a stem cell program) are a pittance compared to the $28.6 billion research budget of the National Institutes of Health, and won't go far to promulgate a national stem cell research agenda.[52] Despite the NIH's insistence that it is solidly behind stem cell research, the funding numbers tell a different story. In 2003, the NIH provided just $27 million for hESC research—only on the 20 or so approved lines—and 8 times that amount for research on adult stem cells. With such out-of-balance funds, it will take substantial muscle from individual states to overcome the federal restrictions. One state pushed back in a very big way.

California declared itself a "restriction-free zone" in late 2002 when then-Governor Gray Davis signed a law that allows nuclear transfer and embryo research. But no money accompanied the legislation. The continued intransigence of the White House forged an alliance between the state senator Deborah Ortiz, Juvenile Diabetes Research Foundation CEO Peter Van Etten, and real estate developer Robert Klein. Ortiz had introduced a $1 billion bond measure on the heels of the Davis legislation, but was unable to move it to a vote. The tall, acerbic Van Etten used his skills as a former Stanford Hospital CEO and his relations with Klein, who sits on the JDRF board, to secure more than $500 million in NIH funding for diabetes research in 2003. For his part, Klein is like many passionate individuals who find themselves at the center of disease advocacy—his 16-year-old son has type I diabetes and his mother suffers from Alzheimer's disease. He became the dynamo behind Proposition 71, a voter initiative on the November

2004 ballot designed to raise bonds to finance stem cell research of every stripe. The bonds would create the California Institute for Regenerative Medicine, presided over by a 29-member board and 3 advisory committees comprised of university officials, stem cell scientists, disease advocacy representatives, and real estate specialists. The proposal contained impressive benefits—and costs. Three billion dollars would be raised and then spent over 15 years primarily in nonprofit California research institutions. The measure would require 30 years and another $3 billion to pay off—a year-over-year expense to California voters of $200 million. Klein sold the measure on the promise that if just one stem cell therapy or affiliated technology is brought to bear on California's $118 billion annual healthcare expenditure, then any criticism of cost will fall by the wayside: a 1 percent savings over several years would easily pay for the entire measure.

The trio knew it needed the endorsement of Hollywood and top scientists to pass the initiative, especially in the thickening atmosphere associated with a presidential election. It signed on Stanford scientists Paul Berg and Irving Weissman (Chapter 7) and University of California, San Diego neuroscientist Hans Kierstead, who had recently announced his promising spinal injury research. In the run-up to the election, the Proposition 71 television campaign churned through $20 million collected from Hollywood elites and Microsoft's Bill Gates, among many others. Klein himself added $2 million to the coffers. California voters saw television spots featuring Michael J. Fox, who suffers from Parkinson's disease, actor Edward James Olmos, and a prerecorded message from Christopher Reeve. Actor Brad Pitt held a news conference announcing his support at a children's hospital in Los Angeles. Berg, Weissman, and University of California, San Francisco, diabetes researcher Jeff Bluestone were featured on the airwaves along with testimonials from patients who expect to benefit from stem cell research. Two weeks before the election, Governor Arnold

Schwarzenegger threw his brawn behind the measure. On November 2, 2004, the California Stem Cell Research and Cures Initiative passed easily by a margin of 59 percent to 41 percent. Klein, a rugged man who looks younger than his 59 years, seems to have his sights set on a national movement that will render federal policy insignificant. "It's no longer a $3 billion success story," he says. "Other states passing legislation will turn this into a $6 billion or even $9 billion crusade."[53]

Three billion dollars is an astounding amount of money, especially when it is focused on a rather small area of biomedical science. The sheer size of the California Institute not only sends a message about the promise of stem cells but also underscores how expensive biomedical research can be and how distant therapies are from human use. Everything begins in the research laboratory, and without the NIH solidly behind a research program, California will replace the NIH and compete alongside other nations who support embryo research and the full spectrum of stem cell biology. The measure—suffering a two-year delay from a lawsuit brought by the religious right—is designed to outlast several presidential administrations and to insulate California from the vagaries of congressional politics. Will California's action shame the federal government into changing its mind? It seems unlikely; in fact, the moral mandate of Bush's second term could lead the White House to retaliate by denying other sources of federal funding to the state. Berg recalls a similar threat made during the recombinant DNA debate of the early 1970s, where failure of a university to adhere to laboratory guidelines put that institution's NIH funding at risk.

Klein is right about others getting into the act. The California initiative set off a frenzy of activity in states who feared that a West Coast magnetized by money would suck the nation's best stem cell biologists right out of their laboratories. Acting New Jersey governor Richard Codey vowed to build a $380 million state-supported Stem Cell Institute, 60 percent of which would be funded by a bond referendum. He has wooed adult stem cell expert Ira Black (Chapter 6) to co-direct the institute along with Rutgers University biologist Wise Young. But Codey is fighting conservatives opposed to the measure, a public leery about the best use of tax dollars, and members of his own democratic party who don't want a divisive issue complicating the November elections. Conservative Missouri politicians repeatedly moved to ban SCNT until finally one of the state's biggest supporters of non-profit science, the Stowers Institute for Medical Research, threatened to withdraw $250 million dollars of its funding for a new Kansas City campus until Missouri law clearly allowed hESC research. At the end of 2005, the institute backed up the warning, giving $6 million to Harvard's rising stem cell star, Kevin Eggan.

The Massachusetts Senate and House voted overwhelmingly in favor of a bill to promote stem cell research in the state while outlawing human reproductive cloning and imposing a series of new regulations on the cutting-edge field. Prior to the vote, Republican governor Mitt Romney threatened to veto the measure if it was approved; with a two-thirds majority margin, he is prevented from doing so. The Illinois State Senate approved legislation that creates the Illinois Regenerative Medicine Institute, to be fueled by a $1 billion general obligation bond issue on the November of 2006 ballot. Similar legislative arm wrestling is going on in other states, including Connecticut, Florida, Kansas, Ohio, Louisiana, Missouri, Maryland, New York, Washington State, and Wisconsin. The Texas legislature slavishly followed Washington, D.C., with competing bills both for and against hESC research. One bill, H.R. 864, makes it a felony—punishable by a fine up to $500,000—for scientists, doctors, and patients to participate in research or medical uses of stem cells made by SCNT. Under the bill, a woman who donates an unfertilized egg to make a cell line for her sick child could face prosecution. Both Texas bills died under the heat of deliberations.

What can Americans expect from unsettled state and federal policy? Some consequences are obvious; others are much less so.

THE BIG CHILL

Early on, the scientific and economic fallout of the administration's policy was swift and immediate. A year after the Bush announcement, a prominent U.S. scientist, Roger Pederson, fled to Britain's Cambridge University, where the laws of Her Majesty's government offer greater scientific freedom. Singapore, eager to develop the first therapeutic advances, sunk $3 billion into biomedical research, built a stem cell institute and aggressively recruited top American scientists and post-doctoral fellows to work for its companies and institutes. Sweden busily commercialized potential therapies. The privately held American company Advanced Cell Therapeutics lost three of its top researchers to Japan and Scotland. Geron, a publicly traded company once valued at more than $1 billion, has had to reduce its scientific staff and has moved some of its stem cell research to Scotland. These events and the predominance of offshore scientific successes signal that America is in the process of losing its chance lead stem cell research. Stanford's Weissman warns, "In 15 years we may be buying drugs derived from stem cell research from China, not the United States."[54]

On the other hand, laboratories like Douglas Melton's show surprising resiliency. Stem cell research at Pederson's former institution, the University of California, San Francisco, continues at a brisk pace and despite his exodus to Europe, some of his research staff stayed behind and joined other investigators. Important technology remains, too—his hESC lines are listed on the NIH registry. Fear of a diaspora of American senior scientists seems to have been, at least in the near term, premature, especially with California holding out the promise of adequate funding. Within leading stem cell laboratories, this worry has been replaced by something much more insidious and far reaching: young biologists are choosing other careers.

No young scientist cares to pin his or her career, the focus of the lab, the futures of graduate students and postdoctoral fellows, and an important research orientation of a department and school on a subject that could be outlawed at any moment. Once labs stop doing research on something, repercussions cascade through the system for years, even decades. In the future—even if policies change—scientific organizations will lack the research infrastructure and trained investigators needed to carry the work forward. "We'll become world experts at mouse stem cell biology," complains Berg, whose own discoveries laid the foundation for scores of new drugs created by the biotechnology industry.[55] Added to the uncertain environment is the discomfort of working on a project that the government bans on the one hand and ignores on the other. Those university investigators with private funding from philanthropy or corporations who chance to work on nonapproved hESCs have set up cloistered sections of their laboratories, with duplicate equipment and signs that say "Entering Non-NIH Space." If a confused graduate student mistakenly carries a cell culture made from a private grant from the room and places it in a government-supplied incubator, the student is potentially breaking the law. It is a bizarre feature in an environment that otherwise enjoys great freedom to operate. A Johns Hopkins University postdoctoral fellow who asked to remain anonymous said, "I do it (hESC research) because I believe what I discover will save lives. But I work in this strange environment, as if I have to sneak around in order to be successful and help people. It's very discouraging."[56]

The repercussions ripple much further than American basic research labs. Pharmaceutical companies are built upon the brains and work of thousands of scientists who receive their training in government-sponsored laboratories. The chief science officers and many founders of biotechnology companies matriculate in the same fashion. In a time of increasing specialization, a decision early in a scientist's career to avoid stem cell research makes it very difficult to return to the field at a later time. In the wake of these decisions, America looses its grip on the entire chain of scientific and commercial events that ensure its discoveries reach its own people.

The disruption to the flow of public information is another reason for worry. The discoveries and experimental results made by scientists show up first in journals, where they can be read, analyzed, and challenged by anyone. The level beauty of publishing an experiment for all to see is the quality control the process imposes on researchers' claims. It begins when the biologist sends the manuscript to the editor. The editor in turn sends it to several domain experts who remain anonymous to the researcher, and they read it critically, examining its data and conclusions. If there is agreement among the group that the science is solid, the editor will publish it. This process of "peer review" has been used for over a century, and as molecular biologist Robert Weinberg put it in an essay about cloning in the *Atlantic Monthly,* "We participate in the peer-review process not only to create a sound edifice of ideas and results for ourselves; we do it for the outside world as well."[57] His point is that if government throttles research on hESCs, the public results that keep everyone informed disappears.

Another argument against the government restrictions is that it drives biological research into the private sector *too early*, skipping the trial-by-fire of peer review. In contrast to their academic counterparts, corporate scientists rely on secrecy to protect their intellectual property from being poached by competitors. And some biotechnology companies, keen to please stockholders and financial analysts, can be too eager to announce their results without having them vetted first in the literature. Weinberg uses American policy on reproductive technologies as a case in point. Because of the congressional bans on embryo research, IVF was forced into private clinics, where it remains an essentially unregulated industry.

UNCERTAIN ECONOMICS

Some say that private money and state initiatives will be enough to move hESC forward. The sheer size of the California project is evidence that private money won't be enough, and what isn't always counted in the financial analysis is money needed from the capital markets. Private money continues to fund hESC research, but at a dangerously low ebb. The problems are caused partly by politics and partly by the residual effects from the 2000 dot-com bust. Early investments give small companies enough staying power to see their inventions reach pre-clinical milestones. Later infusions of cash, including public offerings, enable biotech survivors to reach clinical trials, and partner with drug companies and university laboratories.

Venture capitalists have long said that an uncertain funding environment puts every decision under a microscope. Venture financing lives or dies by investor confidence, and for stem cell companies in particular, that confidence is missing. The message is clear: the problem cannot be solved by California alone. The most profound result of a research ban is that no stem cell companies will form in the first place, because the universe of knowledge will be too small for any investor to take a reasonable risk. Researching the claims of a start-up company means talking to experts in the field, and when the field is constricted by politics, the experts are missing. Where there are no barriers in the pathway from laboratory bench to a patient's bedside, the new technologies and the companies that develop them will flourish.

Collateral damage also occurs from conflicting federal policies. Since 1980, the Bayh-Dole Act has given universities title to inventions made with government funds. By providing universities with the financial incentive to license patents to the private sector, the act was designed as a public service: nonprofits work to identify valuable intellectual property and then, for a fee, license the inventions to companies for development. With very little funding going in, very little intellectual property comes out. With few patents to form the basis of new companies, the domestic capital markets have little to choose from and will invest in other technologies or offshore initiatives.

Bayh-Dole is credited with priming the pump for the biotechnology revolution by offering low-cost licenses of so-called platform technologies—those inventions used by entire industries to make new drugs and therapies for patients—to any company who wants to get into the game. Stifling the creation of new platform technologies made from hESCs—such as drug discovery tools, disease models, or even the lines themselves—means that the first new therapies available to Americans may come from foreign shores, provided of course that using them doesn't mean incarceration.

The world faced a similar quandary 30 years ago amid widespread concerns about the dangers of recombinant DNA research. The Senate came breathtakingly close to criminalizing biomedical research in those days, just as it is threatening to do now. Finally, on the heels of a meeting held by 150 scientists led by Berg and others in a sun-drenched retreat in Asilomar, near Monterey, California, the NIH adopted a framework that included regulation, oversight, and review by the newly formed Recombinant DNA Advisory Committee. This body monitored research that is recognized for transforming science and creating a new industry.

Some predict that stem cell research will eclipse by orders of magnitude the impact of recombinant DNA. What would the world be like if the U.S. government had banned recombinant DNA research back in 1976? Assuming the prohibition persisted during the subsequent decades, and no other research filled the void, the world would be without the Human Genome Project, without the identification of 1,500 disease genes, and without DNA fingerprinting. The U.S. would also be $16 billion poorer in research and investment, $35 billion poorer in revenue, and lacking some 200,000 jobs. Plus, 325 million people would not benefit from over 130 drugs, including vaccines for influenza and hepatitis B.[58]

WHAT DO AMERICANS THINK?

Despite the machinations of federal and state governments, most Americans support stem cell research. In an August 2004 Harris Poll, Americans supported the use of stem cells taken from donated embryos by more than a 6 to 1 margin. The same margin in a similar poll taken in 2001 was 3 to 1.[59] By a margin of 74 percent to 21 percent, Americans (including 79 percent of moderates and 62 percent of conservatives) said in 2004 that they back former first lady Nancy Reagan's call for more stem cell research flexibility.[60] In a February 2005 poll by Results for America, a project of the Civil Society Institute, 70 percent of Americans favored loosening the Bush restrictions. This included more than half of conservatives (56 percent), 80 percent of moderates, and 84 percent of liberals. In a poll conducted a month later by the Coalition for the Advancement of Medical Research, 59 percent of those surveyed favored embryonic stem cell research, and 33 percent were opposed to it.[61]

Whether Americans will voice their approval of stem cell research at a level where it sways the political fortunes of elected officials—or at least their decisions—remains an open question. If treatments using hESCs are discovered and developed elsewhere and used for the benefit of patients in those countries, then American politicians will have to squarely face the patients and families that make up their electorate. It would be an unimaginable act by any politician to deny medical treatment to or imprison sick people for trying to make themselves well. At this moment, political winds in Washington seem to shifting, President Bush's veto threat notwithstanding.

The good news is that embryonic stem cell research, has, as the colloquialism goes, left the barn. Knowledge flows past borders and politicians with speed and ease. Stem cell research will continue despite political prohibitions because people wish it to. What we are left with is whether America will play a major or minor role in its progress—and whether Americans will have access to cures developed where hESC research is permitted. The rate of progress for curing disease with stem cells is a crucially important question; no developed nation has the resources to propel a research agenda like we do—be it a race to the moon or a race to unlock the secrets of the human body.

We face great moral hazards here. The first one is choosing ignorance over knowledge, a dangerous precedent for any society. The second one is allowing politics—and politicians—to intrude on the will of a majority of Americans. Third, and most important, is honoring our obligations to those among us who suffer. Our decisions today will determine whether history regards America at the beginning of the twenty-first century as embarking on a new path of enlightenment or retreating to a dark and pessimistic time.

10
The South Korean Affair

Pride goeth before a fall.
ADAPTED FROM PROVERBS

The medical promise of somatic cell nuclear transfer (SCNT) rests on making human embryonic stem cell (hESC) lines: cloned cells genetically identical to each other and compatible to the human donor.[1] SCNT worked with animal cells, but the first experiments using human cells were disappointing. In 2001, researchers at the biotech company Advanced Cell Therapeutics injected nuclei from human skin cells into an egg without a nucleus—but the embryos failed to develop past the first few cell divisions.[2] In 2003, scientists in China reported fusing human cells taken from various places in the body with enucleated rabbit eggs.[3] The cells removed from inside the growing embryo had the capacity to differentiate into different human cell types, but didn't survive very long. The "species barrier"—the difficulties moving a technology from one animal system to another—was causing researchers the same fits that cloning had twenty years earlier.

The barrier fell in 2004—or so it seemed. A South Korean research team at Seoul National University led by Professor Hwang Woo Suk reported a thriving, immortal hESC line using nuclear transfer; the first of its kind.[4] The announcement caused a major sensation and launched headlines like "Scientists Clone First Human Embryo." True, they had made an embryo, but the real news was the cell line that came from it.

The *Science* paper described how Hwang removed the nuclei of 242 eggs donated by 16 women. A nucleus from a cumulus cell (cumulus cells are developmentally very young and surround the egg in the oviduct) was fused to each empty egg. In each case, both cells came from the same woman. The conditions made blastocysts with an inner cell mass one out of three times. Only one blastocyst produced an hESC line. The South Korean claims seemed reasonable and sound: the hESCs had normal numbers of chromosomes, lived indefinitely, were pluripotent, and formed teratomas, corroborating their embryonic origins. To prove the success of the technique, the cell line had the DNA fingerprint of the donor nucleus, not the eggs.

By itself, the successful application of the technique to human cells was a major advance, thrusting Hwang, a veterinarian from modest roots, into the limelight. But he didn't stop with the first set of experiments. If cloning was possible using human embryonic cells, then why not make therapeutic lines with the nucleus of cells taken from people with disease? In May of 2005, the group reported in a second *Science* paper that they could now derive a so-called "patient-specific" line with as few as 14 eggs, an astonishing increase in the efficiency of their procedure. He claimed over a dozen improvements over pervious methods, including those he perfected during his years of animal research, such as using eggs from women of a certain age and a technique that gently squeezed the nucleus until it popped free from the egg's cell membrane.[5] The second set of experiments had stunning medical implications. A once-distant horizon—using a cell from a patient to generate new healthy cells and tissues that would be transplanted back into the patient without fear of immune rejection—had suddenly catapulted within reach.

If the 2004 paper launched Hwang's career, the 2005 paper put him into orbit. His country treated him like a rock star, lavishing him with nearly $50 million in funding, spanking new laboratories, and lifetime air travel. While most Americans wouldn't recognize a Nobel Prize winner if they sat next to one, Hwang was a man of his people, and regular South Koreans thronged around him, asking for autographs. During his world tours he traveled with an entourage, collected suitcases of humanitarian and scientific awards, and stumped for his World Stem Cell Hub, an array of stem cell banks that would house a collection of custom-made lines for research and therapeutic use. He assured scientists he could teach them the intricacies of his techniques, and arranged funding for his American collaborator, University of Pittsburg's Gerald Schatten. A reproductive biologist with a stem cell center of his own, Schatten was rarely missing from Hwang's side. Schatten served as translator and tour guide, and, according to some researchers, was a clever opportunist. He had good reason to throw his lot in with the South Koreans—he was the only American author on the most important stem cell paper of the year. Due to funding problems at home he feared his lab might be closed and hoped for a soft landing after Hwang tapped him to lead the Western emirates of the world hub.

In the summer of 2005, cracks began to appear in the South Korean story. A news story in the journal *Nature* suggested female associates in Hwang's lab supplied eggs for the research. This caused an ethical stir among Westerners, raising questions about whether the women were coerced into donating their eggs. After reviewing consent documents connected to the study, two bioethicists at Stanford University, David Magnus and Mildred Cho, suggested that the women hadn't been informed fully of the medical risks.[6] Then, a South Korean TV news program, *PD Notebook*, received a tip from a former collaborator of Hwang who claimed that compared to the first study, the 2005 results looked suspicious. *PD Notebook* relentlessly dogged the allegation, airing information received from other whistle-blowers and investigative journalists. In November Hwang went public, admitting that his female lab workers had donated eggs and had been paid for them. He denied knowing about it.

The increasing attention caught the eye of young scientists inside Korea who frequented Internet chat rooms, particularly a website called BRIC, short for Biological Research Information Center. Pages of supplemental data supporting the experiments were available on *Science*'s website, and soon anonymous postings on BRIC reported major inconsistencies. Not only did photos of supposedly different cell cultures in the 2005 paper appear to be the same, the crucial DNA fingerprinting data was suspect. The publicity compelled Hwang to retest the samples, and after doing so, he couldn't get his results to jibe. He claimed his collaborators at the fertility clinic had switched the eggs. He asked *Science* to retract the 2005 paper. Pittsburgh's Schatten announced he was washing his hands of the matter, severing his ties to the man whose fortunes he had followed. The wheels had begun to come off the Hwang wagon.

Hwang's employer, Seoul National University (SNU), announced they would investigate, and it didn't take long for the panel to issue its report. On January 10, 2006, the other shoe dropped: both papers were faked. The 2004 line had been made from a frozen embryo, not from nuclear transfer. Of the 11 lines claimed in the 2005 paper, only two actually existed, and both came from frozen embryos. An analysis of a dog, Snuppy, which Hwang reported in the journal *Nature*, was revealed to be a true clone. Most disturbing was the report's mention of the human eggs procured by the lab over a three-year period. An astounding 2,061 eggs from 129 women had been used, and laboratory journals seized from the lab listed the use of 283 eggs, not 185, for the

2005 paper.[7] Government investigators raided Hwang's home, and women's rights groups sued the government after news broke that Hwang had escorted donors to the clinic. In tears, Hwang admitted he lied about the egg donations. The government launched an audit into SNU's use of research funds. Then, SNU fired Hwang.

Stateside, *Science* quickly retracted the second paper. The disaster swept over Schatten, too. After its own probe, the University of Pittsburg found the biologist guilty of "research misbehavior," saying as corresponding author he was derelict in his duty to ensure the data in the 2005 paper was solid and aboveboard. The panel also questioned his evasive behavior and his judgment in accepting $40,000 in honoraria, including a $10,000 cash payment slipped to him at a news conference. Two months later, the university itself came under fire from a Pittsburg newspaper. Reporters found a university oversight committee skipped a full review of Schatten's research after he assured them it did not involve egg donors he could identify. Federal regulation requires investigators to provide written confirmation that they have no knowledge of who the donors are; the university didn't ask for the document and Schatten didn't offer it. Even *Science* didn't escape criticism from those who wondered how the reviewers and editors of a prestigious journal could miss such a stunning set of lies.

When the dust settled, the Hwang scandal went down as the worst scientific fraud in recent memory. Destined for history books, it featured a basketful of misdeeds, ranging from shameless falsehoods to venal sins. Other facts surfaced too, including the complicated network of responsibilities among a large number of collaborators; there were 25 authors listed on the 2005 paper. The assembly line that Hwang set in motion made it difficult for researchers and technicians to grasp the downstream consequences of their actions or to question shaky experimental results. The stern hierarchy of the South Korean's laboratory, coupled with pressure to perform under an international spotlight, exposed how junior researchers perpetuated the deceit and became fearful of speaking out. The differences in ethical norms between nations surfaced, too. During the first weeks of the scandal, South Koreans bristled at Western commentators who suggested its scientists were unaware of international ethical standards for biomedical research, despite a research study in South Korea that showed otherwise.[8]

Others wondered whether the usually reliable mechanism of scientific peer-review needed an overhaul. The 2005 paper was accepted more swiftly than most, probably because it claimed to improve upon the 2004 experiments and because journals strive to publish important results, and publish them quickly. The editor of *Science,* Donald Kennedy, said that while only two papers had been retracted because of scientific misconduct during his five years at the magazine, he would assemble a group of outside experts to help the journal's editors dissect the review process and search for ways to improve it. Time will tell whether improvements can be made; veteran peer-reviewers say that it is difficult to catch a skillful liar, especially one who claims to have mastered the arcane nuance of a new technique.

The scandal also raised the awareness of the health complications associated with donating eggs, the same procedure used in *in vitro* fertilization clinics. Women must first take daily injections of hormones over several weeks, stimulating the release as much as a dozen or more eggs. Then under anesthesia, a vaginally inserted probe guides a needle to the ovaries, and retrieves the eggs with light suction. Each step in the procedure carries risks. Egg retrieval is invasive and uncomfortable, and in about five percent of women, the hormones cause ovarian hyperstimulation syndrome (OHSS). In rare cases, OHSS is severe enough to require hospitalization. Sixteen of 129 South Korean donors required medical treatment for OHSS, and two were hospitalized.

In the end, the South Korean affair laid bare the enormity of what is at stake for embryonic stem cell research. Hwang, it seems, desperately wanted to prove that his facility with animal cells could work with human lines. A success would put him, his little-known laboratory, and his industrious nation at center stage. Others cheered for him, too, for his feat meant that embryonic stem cell research was one step closer to a therapeutic reality.

Measured by the timescale of scientific progress, this storm passed quickly. Other labs set about testing the Hwang method and devising new techniques of nuclear transfer, just as they would have if his experiments were legitimate. Scientists are born skeptics, and the proof of the pudding is whether an experiment is repeatable, and whether it spawns new lines of inquiry and useful discoveries. The notion of repeatability begs the biggest question connected to the fraud: If Hwang knew his data was faked, why was he so cavalier about it? Why accept awards, travel around the world, and give speeches about how his discovery would transform medicine? Hwang was a veteran researcher—a successful animal cloner—who surely knew that sooner or later others, trying to repeat his results, would find him out.

Only Hwang Woo Suk knows the answers to these questions. But it hasn't stopped speculation about his motives. The most charitable view maintains his fame and globetrotting ways blinded him to the problems at home—the inmates were running the asylum. Given the import of his claims, his style of micromanagement, and his reputation for staying up all night in the lab, this scenario seems unlikely. Others point out that a combination of hubris, frustration with failed experiments, and pressure from his government led him to cook his books. Another explanation suggests he knew he was missing a small but elemental piece of the puzzle, and figured once he published a theoretical—and falsified—set of protocols many labs would jump on the chase. Once someone else found the elusive bit, he would successfully repeat the experiment and claim victory, saying different roads could lead to the same result.

In the end, what Hwang thought doesn't really matter. But his misdeeds caused three casualties. The debacle tarnished the reputation of science and scientists, who rely on a bedrock principle, honesty. Without honest researchers and the faith of institutions such as universities, scientific journals, and the communities that back them, the notion of research for the public good is little more than an empty promise. Though scientific fraud has always been with us, this fact is of little solace when we face it so squarely. The second consequence springs from our government's policy. Years passed since the first nuclear transfer paper, time that could have been used by American researchers with unfettered access to funding for embryonic stem cell research. It is conceivable that many scientists working in the world's most technologically advanced laboratories could have begun the process of unraveling the knot of lies perpetuated by the South Korean. Instead, Americans stood on the sidelines. Finally, and most importantly, the deceit dashed the hope of millions of suffering people and their families.

The lesson for scientists everywhere is that though fame, fortune, and the answers to their most precious questions may fuel their quests, these achievements pale in comparison to what is at stake for persons who stand to benefit from their discoveries. With this in mind, the vast collective of biomedical science moves toward honest and beneficent ends. It is only a matter of time before real patient-specific lines of embryonic stem cells are made, taking the next step along a road that will reveal a thousand new facets of human biology and how to help the infirm and sick among us.

Epilogue

The more a scientific field deals with human affairs, the greater the chance that scientific theories will clash with traditions and beliefs.[1]

FRANCOIS JACOB

On the day that Ronald Reagan died of complications from Alzheimer's disease, I was attending a conference organized by the Stem Cell Action Network, an advocacy group representing patients with Alzheimer's, multiple sclerosis, spinal cord injury, diabetes, and liver, kidney, and heart ailments. Most of the hundreds in attendance were family members of people suffering from chronic disease or disability. A fair number of disabled persons were there—some in wheelchairs. At my table were the parents of a paralyzed teenager, the father of a child with juvenile diabetes, and a young woman whose best friend is paralyzed from the waist down. Stem cell scientists, ethicists, and representatives of state and national initiatives promoting stem cell research took the podium in turn and addressed the audience. One speaker, an internationally renowned stem cell biologist, made the point that it was too early to think about cures. Between presentations, we introduced ourselves and chatted about our reasons for coming. It became clear that the caution expressed at the podium was lost on many in the audience. As the scientists and ethicists talked on, the group at my table grew restive.

Then came a video of a rat with a crushed spinal cord: the same experiment featured in Chapter 7. The disabled rat walked again after a transplant with nerve cells made from a line of embryonic stem cells. The video brought people to the edge of their seats. The questions that the caregivers asked betrayed a singular, and remarkable, focus: "All this discussion about science and ethics is interesting and important, but when will this help my son, my wife, my friend?"

After the first round of talks, the parents of the paralyzed teenager pushed a dog-eared file folder across the table and asked me to look at its contents. Most of the material had been pulled from the Internet. The headlines announced "imminent cures" and "groundbreaking results." But what caught my attention was a news article, the text highlighted with a fluorescent marker.

A clinical trial in Portugal had just begun using cell therapy to treat spinal cord injury. The procedure seemed bizarre: cells removed from inside the nose were transplanted into the injured spinal cords of paralyzed patients. The doctors first removed a one-inch section of mucosal tissue from the roof of the nasal cavity, where millions of specialized nerve cells carry sensory information from the nose through the olfactory nerve to the brain. The olfactory nerve is the only cranial nerve that regenerates—a special cell called an OEG (olfactory ensheathing glial) cell is thought to contribute to the nerve's regeneration. Then surgeons cut out a damaged section of the spinal cord and pressed the mucosal tissue into the incision. The article reported that the first dozen or so patients undergoing the procedure had improved. Most had regained a degree of muscle control and sensation. The procedure was not covered by insurance and was horribly expensive, costing upwards of $50,000.

I asked the parents whether they had researched the peer-reviewed medical literature for the results. They said they had not. I asked if they planned to take their teenager to Portugal, and they replied yes, probably. The young woman at the table said that she too had heard of the Portuguese study and was disappointed that none of the speakers mentioned the promising results. It didn't matter that the famous scientists in the room were warning that stem cell transplants at this stage in the game are a premature and risky business. If there was hope to be had, people would grasp it without question.

Some months later, I looked into the Portuguese clinical trials. A familiar problem had emerged. The trials were immediately politicized. Opponents of embryonic stem cell research played up the Portuguese study on their websites, citing the trials as yet another reason why embryonic stem cells are unnecessary. Here, American scientists urged caution. Rutgers University neuroscientist and surgeon, Wise Young, who was Christopher Reeve's doctor, went on record saying that although he supported such research in patients with no other options, he had serious reservations about the European studies.[2] The clinical data had not been published, and like critics of the heart disease trials using unpurified bone marrow, Young said transplanting mixed populations of cells left scientists in the dark about what exactly was causing the improvement. Also he was aghast that the Portuguese surgeon removed a part of the spinal cord, a procedure he considered unusually risky.

The offshore human experiments continued. A year later, over 50 patients had enrolled in the Portuguese trial, including several Americans. According to press reports, only four experienced temporary side effects; most showed improvement.[3] Then, China began a similar trial. In less than a year, they had treated hundreds of patients. In Australia, a similar clinical trial using mucosal cells has also started.[4] In early 2005, a Detroit spinal injury clinic opened to help American patients enroll in the Chinese and Portuguese trials. As of this writing, neither trial had published the data from the surgeries, and both remain controversial.

The parents at the conference could not wait for months or years. Had they taken their child to Portugal or China and spent $50,000 on a risky procedure? Had the U.S. permitted embryonic stem cell research years earlier, would spinal injury trials already be approved for American patients? Which approach is best here—march right in or go safe and slow? Is the Federal Drug Administration being too conservative in delaying domestic human trials? Has our debate over embryonic stem cells become so fractious that it has paralyzed our research community?

The answers are hard to find. Yet it *is* clear that the embryo "proxy" war—as some politicians and commentators call the conflict encompassing the issues of abortion, *in vitro* fertilization, and research using embryos and embryonic stem cells—has exacted a toll on American science and medicine. We have ceded leadership to others. At great expense and risk to their health, sick Americans travel overseas for new therapies unavailable at home. While opposing groups wrangle, time passes and hope withers for those who suffer. I believe it is impossible to harm a blastocyst. These 100 cells are not conscious of existence and do not feel. The scant beginnings of a nervous system are a week away. During the first month of development, there is no behavioral evidence of neural function, and reflex activity isn't detected until six weeks after fertilization. The first neural circuitry, where a sensory impulse can be forwarded to the brain through the spinal cord, is 11 weeks distant.[5]

What about the human potential of these cells? I think about this question in terms of potential persons and actual persons. It is hard for me to imagine how we should treat persons who don't yet exist. Neuroscientist and presidential advisor Michael Gazzaniga contends that potential is not innate to the blastocyst. "There's potential for 30 homes in a Home Depot, but if the Home Depot burns down, the headline isn't '30 Homes Burn Down.' It's 'Home Depot Burns Down,'" he says.[6] Theologian Ernlé Young expresses the logic another way: "Is potentiality the same as actuality? You and I are potentially dead. But we are not yet dead, and until we are actually dead, it would be a mistake to treat us as if we were dead."[7]

Does this mean that reverence and respect for the blastocyst are not important? Making or using blastocysts for no humanitarian purpose or to sell them on the open market is unequivocally wrong. Experimenting with them in order to create a human being is unquestionably wrong.

The *Brave New World* view argues that we stand at the precipice of a slippery slope. It contends that using blastocysts for research amounts to human commoditization, that it inevitably encourages a marketplace of embryo farms, fetuses made for spare parts and cloned human beings. I think this scenario is unlikely, based on our history.

The steady march of humankind's medical discoveries has been overwhelmingly used for good, not for evil. In the developed world, we live longer lives and enjoy unprecedented health because of advances in medicine. In hospitals and clinics, our ethics committees ensure that patients are treated with respect and dignity. The American drug and medical device industries are among the most heavily regulated in the world. To say that medical science has an unblemished record is, of course, wrong. There are instances of abusive and illegal human experimentation, cases where medical professionals have intentionally or negligently hurt patients entrusted to their care, and examples where drugs were too hastily approved for human use. When such things happen, we strive mightily to erect ever more rigorous safeguards. Yet even these blemishes do not diminish our history of using research and medicine to improve our lot in life.

Some say that biomedical science moves too quickly; it intrudes too deeply into the natural world, and sooner or later, there will be no mysteries left to solve. On that day, we will lose our innocence, and perhaps our humanity. However, there has never been a shortage of awe-inspiring challenges. Every time we uncover a mystery, another awaits us. We pursue knowledge about biology and our relations to nature as naturally as we breathe. We do so because our acts and efforts generate hope—hope for legions of parents, children, husbands, wives, and friends who need those cures. The optimism that we can improve life and relieve suffering *is* our humanity.

The people who came to the meeting one year ago—and those who attend hundreds of meetings just like it all over the country—come for a reason more important than science, ethics, or politics. They come for hope. Is there hope for their child, their parent, or their friend? Will they find it today or years from now? Will they find it here, in America, or will they have to go elsewhere?

Notes

CHAPTER 1

1. Harold Varmus. Congressional testimony. December 1998. http://www.hhs.gov/asl/testify/t981202a.html.
2. James A. Thomson, et al. "Embryonic stem cell lines derived from human blastocysts." *Science*, 282(5391) (1998):1145-1147.
3. In the end, the British research didn't work because the researchers attempting the derivations had no experience with primate cells, which behaved differently in the laboratory than mouse embryonic stem cells.
4. James Thomson. Interview with author. May 2004.
5. James A. Thomson, et al. "Isolation of a primate embryonic stem cell line." *Proceedings of the National Academies of Science USA*, 92 (1995):7844-7848.
6. Ted Golos. Interview with author. May 2004.
7. http://www.cnn.com/HEALTH/9811/05/stem.cell.discovery/.
8. Richard M. Doerflinger. "The threat of science without humanity." http://www.nrlc.org/news/1999/NRL899/doer.html.
9. The Pope's remarks can be found at http://www.nrlc.org/news/1999/NRL1099/stem.html.

10. CNN news. August 9, 2001. "Candidate Bush opposed embryo stem cell research." http://archives.cnn.com/2001/ALLPOLITICS/08/09/bush.history.stem.cell/.
11. AP press. May 20, 2005. "Bush vows to veto stem cell legislation." http://www.ap.org/.

CHAPTER 3

1. Thomas Bartman, et al. "Early myocardial function affects endocardial cushion development in zebrafish." *Public Library of Science Biology*, 2(5) (2004):e129.
2. Matthew Scott. This statement and following statements from interview with author. June 2004.

CHAPTER 4

1. See Ian Wilmut's interview one year after the birth of Dolly at the Academy of Achievement's website: http://www.achievement.org/autodoc/page/wil0int-5.
2. A slightly dramatized opening scene. For a history of Leroy Steven's discoveries, see the Jackson Laboratory corporate website: http://www.jax.org/mission/tjl_advances.html, and Ricki Lewis. "A stem cell legacy: the trail begins with cigarettes and strain 129 mice." *The Scientist*, 14 (5) (March 6, 2000):19.
3. Jackson Laboratory is the world's largest mammalian genetics research center; home of 1,300 scientists and more than 2 million laboratory mice, including many special varieties that emulate human disease.
4. Not surprisingly, human and mouse tumors arise in much the same way. Malignant forms of teratomas, or teratocarcinomas, are one of the most common (and treatable) cancers in young men,

including seven-time Tour de France winner Lance Armstrong. Mr. Armstrong's testicular tumor was small, but the cancer had spread to his lungs and brain, threatening his life. His cancer was treated successfully with chemotherapy. Benign teratomas are found more commonly in women. In this case, an unfertilized egg prematurely begins embryogenesis. The resulting structures are called dermoid cysts and they later develop into teratomas.

5. Gail R. Martin. "Isolation of a pluripotent cell line from early mouse embryos cultured in medium conditioned by teratocarcinoma stem cells." *Proceedings of the National Academy of Sciences USA*, 78 (1981):7634-7638.
6. There is no Swiss cheese-shaped Nobel Prize for the lab mouse, but there should be: more than 17 Nobel laureates credit the small rodent for their discoveries. Its genome was published in 2002 before the human genome and over 30,000,000 mice soldier on every year, helping researchers develop treatments for diseases such as Alzheimer's, AIDS, cystic fibrosis, cancer, diabetes, and glaucoma.
7. Robert Briggs and Thomas J. King. "Transplantation of living nuclei from blastula cells into enucleated frog's eggs." *Proceedings of the National Academy of Sciences USA*, 38 (1952):455-463.
8. Robert Briggs and Thomas J. King. "Changes in the nuclei of differentiating endoderm cells as revealed by nuclear transplantation." *Journal of Embryology and Experimental Morphology*, 100 (1967):269-312.
9. John Gurdon, et al. "The developmental capacity of nuclei transplanted from keratinized cells of adult frogs." *Journal of Embryology and Experimental Morphology*, 34 (1975):93-112.
10. Karl Illmensee and Peter C. Hoppe. "Nuclear transplantation in Mus musculus: developmental potential of nuclei from preimplantation embryos." *Cell*, 23 (1981):9-18.

11. James McGrath and Davor Solter. "Inability of mouse blastomeres transferred to enucleated zygotes to support development in vitro." *Science*, 226 (1984):1317-1319.
12. Willadsen did not publish his early results, but his nuclear transfer techniques were repeated in cattle. See Steen Willadsen et al., "The variability of late morulae and blastocysts produced by nuclear transplantation in cattle." *Theriogenology*, 35 (1991):161-170.
13. Ian Wilmut, et al. "Viable offspring derived from fetal and adult mammalian cells." *Nature*, 385 (1997):810-813.
14. Wilmut reported the birth of lambs, named Megan and Moran, following nuclear transfer from established tissue-culture cells a year earlier using a cell line derived from a 9-day-old embryo. See K. H. S. Campbell, et al. "Sheep cloned by nuclear transfer from a cultured cell line," *Nature*, 380 (1996):64-66.
15. Rudolph Jaenisch. Interview with author. July 2004.
16. Irving L. Weissman, et al. "Scientific and medical aspects of human reproductive cloning." National Academy of Sciences Panel on Scientific and Medical Aspects of Human Cloning, 2002.
17. Ian Wilmut, et al. "Human cloning: can it be made safe?," *Nature Reviews*, 4 (2003):855-864.
18. A similar condition, Beckwith-Wiedman syndrome, is found in humans. Its incidence increases in children born by in vitro fertilization.
19. See cloning statistics at http://www.fda.gov/fdac/features/2003/303_clone.html.
20. See note 17.
21. D. P. Wolf. "Assisted reproductive technologies in rhesus macaques." *Reproductive Biology and Endocrinology*, 2(37) (2004):6.
22. Philip Cohen. "Human reproductive cloning 'currently impossible'" (April 10, 2003). NewScientist.com news service. http://www.newscientist.com/article.ns?id=dn3614.

23. The International Society of Stem Cell Research (ISSCR) prefers the term nuclear transfer to the term therapeutic cloning. On the ISSCR website, it explains: "Cells generated by nuclear transfer are by no definition a clone of the donor of the transferred nucleus, as the enucleated oocyte still contains numerous stable maternal proteins that are transmitted for several cellular divisions and could impact on gene expression, numerous types of maternal RNA, as well as maternal mitochondrial DNA. The term 'therapeutic cloning' as used to describe all stem cells generated through nuclear transfer is therefore scientifically erroneous and misleading. The term 'therapeutic' is equally misleading. While the stem cell research community is dedicated to the search for therapies using the technique of nuclear transfer, it is far too early to predict therapeutic uses, and naming a technique for its hoped outcome may inadvertently offer premature hope to desperate patients and families." See the ISSCR website at http://www.isscr.org/.
24. Melissa K. Carpenter, et al. "Protocols for the isolation and maintenance of human embryonic stem cells." *Human Embryonic Stem Cells* (Totowa NJ: Humana Press, A.Y. Chiu and M. S. Rao, eds., 2003), 116-120.
25. Lines of other cell types have been long used as experimental tools and can diagnose and treat disease. Immortal lines of white blood cells called hybridomas produce pure concentrations of antibodies, a kind of protein that can be used therapeutically to control rejection of an organ transplant. Common pregnancy or diabetes tests use antibodies derived from hybridomas to identify proteins in the urine or glucose levels in the blood. The most prolific cell lines come from the blood and tissues of cancer patients. Thousands of cell lines representing many types of cancer have been established in order to study how the disease progresses and how drugs can affect its outcome.

26. Chunhui Xu, Melissa K. Carpenter, et al. "Feeder-free growth of undifferentiated human ESC." *Nature Biotechnology*, 19 (2001): 971-974.
27. K. Hsaio, et al. "Correlative memory deficits, a elevation, and amyloid plaques in transgenic mice." *Science*, 274(5284) (1996):99-103.
28. Ronald B. DeMattos. "Apolipoprotein E dose-dependent modulation of beta-amyloid deposition in a transgenic mouse model of Alzheimer's disease." *Journal of Molecular Neuroscience*, 23(3) (2004):255-62.

CHAPTER 5

1. The totipotence prize goes to the plant kingdom. Any cell—stem, leaf, or root—can in principle develop into a mature plant. Cloning identical copies of plants using tissue culture has revolutionized whole sectors of agriculture. Generally, the older the plant, the harder it is to clone.
2. An exception to the all-cells-have-the-same-genes rule is found in the immune system. In vertebrates, selective loss of DNA gives these cells the ability to respond to a bewildering array of foreign invaders and infectious agents. Recent evidence suggests that our ability to distinguish different smells works the same way (see Chapter 6).
3. Stuart Kim, et al. "Differentiation of insulin-producing cells from human neural progenitor cells." *Public Library of Science Medicine*, 2(4) (2005):e103.
4. Nicholas R. Forsyth, et al. "Telomerase and differentiation in multicellular organisms: turn it off, turn it on, and turn it off again." *Differentiation*, 69 (2002):188-197.
5. Martin Raff. "Adult stem cell plasticity: fact or artifact?," *Annual Review of Cell Developmental Biology*, 19 (2003):1-22.
6. Ibid., 6.

7. Nobuko Uchida, et al. "Direct isolation of human central nervous system stem cells." *Proceedings of the National Academy of Sciences USA*, 97 (2000):14720-14725.
8. Houman D. Hemmati, et al. "Cancerous stem cells can arise from pediatric brain tumors." *Proceedings of the National Academies of Sciences USA*, 100(25) (2003):15178-15183.
9. Joaquin Cortiella. "Tissue engineered lung using somatic lung progenitor cells." Abstract, International Society for Stem Cell Research Annual Meeting, June 10-13, 2004.
10. Karl-Ludwig Laugwitz, et al. "Postnatal isl1+ cardioblasts enter fully differentiated cardiomyocyte lineages." *Nature*, 433 (2005):647-655.
11. Hanna Mikkola. "Placenta as a novel site for hematopoietic stem cells." Abstract, International Society for Stem Cell Research Annual Meeting, June 10-13, 2004.
12. Allan Spradling. "More ovarian stem cells." Abstract, International Society for Stem Cell Research Annual Meeting, June 10-13, 2004.
13. There are over 150 CD cell surface markers.

CHAPTER 6

1. Emphasis (!) added. From Reuters News Report, March 24, 2005. http://reuters.com/newsArticle.jhtml?type=healthNews&storyID=7991470.
2. From WebMD. http://my.webmd.com/content/article/82/97161.
3. From the Senior Journal. http://www.seniorjournal.com.
4. John D. Gearhart, et al. "Derivation of pluripotent stem cells from cultured human primordial germ cells." *Proceedings of the National Academy of Sciences USA*, 95(23) (1998):13726–13731.

5. Kaomei Guan, et al. "Pluripotency of spermatogonial stem cells from adult mouse testis." *Nature,* 440 (2006):1199–1203.
6. Anders Bjorklund, et al. "Embryonic stem cells develop into functional dopaminergic neurons after transplantation into a Parkinson's rat model." *Proceedings of the National Academy of Sciences*, 99 (2002):2344-2349.
7. Jong-Hoon Kim and Ron McKay. "Dopamine neurons derived from embryonic stem cells function in an animal model of Parkinson's disease." *Nature*, 418 (2002):50-55.
8. Benjamin Reubinoff, et al. "Human embryonic stem cell-derived neural progenitors correct behavioral deficits in Parkinsonian rats." Abstract, International Society for Stem Cell Research Annual Meeting, June 10-13, 2004.
9. Rudolph Jaenisch. Interview with author. September 2004.
10. Konrad Hochedlinger, et al. "Reprogramming of a melanoma genome by nuclear transplantation." *Genes and Development*, 18 (2004):1875-1885.
11. William Rideout, et al. "Correction of a genetic defect by nuclear transplantation and combined cell and gene therapy." *Cell*, 109 (2002):17-27.
12. For information and statistics on diabetes, see the International Diabetes Federation. http://www.idf.org/home/index.cfm?node=264, or the Juvenile Diabetes Research Foundation. http://www.jdrf.org/.
13. Yuval Dor, et al. "Adult pancreatic β-cells are formed by self-duplication rather than stem-cell differentiation." *Nature*, 429 (2004):41-46.
14. For an excellent review of this paper, see Ken Zaret. "Self-help for insulin cells." *Nature*, 429 (2004):30-31.
15. Yuehua Jiang, et al. "Pluripotency of mensenchymal stem cells derived from adult marrow." *Nature*, 418 (2002):41-49. Hong Ye Jiang, et al. "Multipotent progenitor cells can be isolated from post-

natal murine bone marrow, muscle, and brain." *Experimental Hematology*, 30 (2002):896-904.

16. Rebecca J. Morris, et al. "Capturing and profiling adult hair follicle stem cells." *Nature Biotechnology*, 4 (2004):393-394.
17. Kursad Turksen and Tammy-Claire Troy. "ES cell differentiation into the hair follicle lineage in vitro." *Methods in Molecular Biology*, 185 (2002):255-260.
18. Tudorita Tumbar, et al. "Defining the epithelial stem cell niche in skin." *Science*, 303 (2004):359-363.
19. Elaine Fuchs, et al. "Socializing with the neighbors: stem cells and their niche." *Cell*, 116 (2004):769-778.
20. Stewart Sell. "Adult stem cell plasticity: introduction to the first issue of Stem Cell Reviews." *Stem Cell Reviews*, 1 (2005):1-5.
21. Amy Wagers. Interview with author. May 2005.
22. Jonas Mattsson, et al. "Lung epithelial cells and type II pneumocytes of donor origin after allogeneic hematopoietic stem cell transplantation." *Transplantation*, 78(2004):154-157.
23. Guillermo Munoz-Elias, et al. "Adult bone marrow stromal cells in the embryonic brain: engraftment, migration, differentiation, and long term survival." *Journal of Neuroscience*, 24 (2004):4585-4595.
24. Diane Krause, et al. "Lack of a fusion requirement for development of bone marrow-derived epithelia." *Science*, 305 (2004):90-93.
25. Diane Krause. Interview with author. May 2005.
26. Jackie R. Bickenbach and Matthew M. Stern. "Plasticity of epidermal cells: survival in various environments." *Stem Cell Reviews*, 1 (2005):71-77.
27. Alan Trounson. "Stem cells, plasticity and cancer—uncomfortable bedfellows." *Development*, 131 (2004):2763-2768.
28. See note 20.
29. Amy Wagers, et al. "Little evidence for developmental plasticity of adult hematopoietic stem cells." *Science*, 297 (2002):2256-2259.

30. Amy Wagers and Irving Weissman. "Plasticity of adult stem cells." *Cell*, 116 (2004):639-648.

31. Craig Dorrell and Markus Grompe. "Liver repair by intra- and extrahepatic progenitors." *Stem Cell Reviews*, 1 (2005):60-64.

32. Helen Blau, et al. "Hematopoietic contribution to skeletal muscle regeneration by myelomonocytic precursors." *Proceedings of the National Academies of Science USA*, 101 (2004):13507-13512.

CHAPTER 7

1. See Christopher Reeve's 2003 interview with Jerome Groopman. http://www.jeromegroopman.com/reeve.html.

2. See the International Society for Heart and Lung Transplantation for current statistics. http://www.ishlt.org.

3. Anthony Atala and Chester Koh. "Tissue engineering applications of therapeutic cloning." *Annual Review of Biomedical Engineering*, 6 (2004):27-40.

4. See the National Diabetes Information Clearinghouse. http://diabetes.niddk.nih.gov/.

5. American Cancer Society. "Cancer facts and figures 2004." http://www.cancer.org/.

6. See the American Heart Association's Heart Disease and Stroke Statistics—2004 Update. http://www.americanheart.org.

7. Arlene Y. Chiu and Mahendra S. Rao. "Human embryonic vs. adult stem cells." *Human Embryonic Stem Cells* (Totowa, NJ: Humana Press, Chiu and Rao, eds., 2003), 239-257.

8. Irving L. Weissman. "Stem cells: lessons from the past, lessons for the future." Vatican Science Counsel Report, unpublished (2004):13.

9. Frederick Appelbaum. "Hematopoietic cell transplantation as immunotherapy." *Nature*, 411 (2001):385-389.

10. Emil Frei and Emil Jay Freireich. "Progress and perspectives in the chemotherapy of acute leukemia." *Advances in Chemotherapy*, 2(1965):269-298.
11. Nobuko Uchida, et al. "High doses of purified stem cells cause early hematopoietic recovery in syngeneic and allogeneic hosts." *Journal of Clinical Investigation*, 101 (1998):961-968.
12. Ewa Carrier and Gracy Ledingham. *100 Questions and Answers about Bone Marrow and Stem Cell Transplantation* (Boston, MA: Jones and Bartlett, 2004), 14.
13. Data obtained from Bertram Lubin, M.D., President, Children's Hospital Research Institute, Oakland. Interview with author. March 2005.
14. Irving Weissman. Interview with author. August 2004. According to Weissman, he did the first experiment to test this idea in 1957 in Great Falls, Montana. "It was a high school biology project," he recalls. "I irradiated a female mouse and then transplanted the bone marrow from the male mouse to the female. A short time later, I transferred a patch of skin from the same male mouse to the female. Normally, skin would not take. But there was no rejection."
15. Donald Orlic, et al. "Bone marrow cells regenerate infarcted myocardium." *Nature*, 410 (2001):701-705.
16. Nicholas Wade. "Tracking the uncertain science of growing heart cells." *The New York Times* (March 14, 2005):p. A1.
17. Christof Stamm, et al. "Autologous bone marrow stem cell transplantation for myocardial regeneration." *Lancet*, 361 (2003):45-46.
18. Anthony Mathur and J. F. Martin. "Stem cells and repair of the heart." *Lancet*, 364 (2004):183-192.
19. Leora B. Balsam, et al. "Hematopoietic stem cells adopt mature hematopoietic fates in ischemic myocardium." *Nature*, 428 (2004):668-673.

20. Kai C. Wollert, et al. "Intracoronary autologous bone-marrow cell transfer after myocardial infarction: the BOOST randomised controlled clinical trial." *Lancet*, 364 (2004):141-148.
21. Robert Lanza, et al. "Regeneration of the infarcted heart with stem cells derived by nuclear tranplantation." *Circulation Research*, 94 (2004):820.
22. Lanza also reports that ACT has fabricated a three-dimensional "miniature heart" by using dog stem cells. He envisions human versions of such structures used as "natural" pacemakers.
23. Amy Wagers. Interview with author. May 2005.
24. Kenneth Chien. "Lost in translation." *Nature*, 428 (2005):607-608.
25. Karl-Ludwig Laugwitz, et al. "Postnatal isl1+ cardioblasts enter fully differentiated cardiomyocyte lineages." *Nature*, 433 (2005):647-655.
26. Joshua Hare. Interview with author. June 2005.
27. See note 9, Chapter 5.
28. Shulamit Levenberg, et al. "Differentiation of human embryonic stem cells on three dimensional polymer scaffolds." Abstract, International Society for Stem Cell Research Annual Meeting, June 10-13, 2004.
29. Anthony Atala. This statement and following statements from interview with author. October 2004.
30. Hans S. Keirstead, et al. "Oligodendrocyte progenitor transplants and spinal cord injury." *Journal of Neuroscience*, 25(19) (2005): 4694-4705.
31. Anders Bjorklund. Interview with author. September 2004.
32. O. Lindval, et al. "Fetal dopamine-rich mesencephalic grafts in Parkinson's disease." *Lancet*, 2 (1988):1483-4.
33. See the NIH news release. "Fetal cell therapy benefits some Parkinson's patients: first controlled clinical trial shows mixed results." http://www.nih.gov/news/pr/apr99/ninds-21.htm.

34. Curt R. Freed, et al. "Transplantation of embryonic dopamine neurons for severe Parkinson's disease." *The New England Journal of Medicine*, 344 (2001):710-719.
35. Ann Tsukamoto. Interview with author. May 2005.
36. Alan Trounson. Interview with author. November 2004.
37. See note 4, Chapter 1.
38. Frank Marini, et al. "Mesenchymal stem cells: potential precursors for tumor stroma and targeted-delivery vehicles for anticancer agents." *Journal of the National Cancer Institute*, 96 (2004):1593-1603.
39. See note 31.
40. Akira Nakamizo, et al. "Human bone marrow-derived mesenchymal stem cells in the treatment of gliomas." *Cancer Research*, 65(8):3307-18.
41. Yuri Verlinsky. "Preimplantation genetic diagnosis as a source of human embryonic stem cell lines." Abstract, International Society for Stem Cell Research Annual Meeting, June 10-13, 2004.

CHAPTER 8

1. Thomas Huxley. *Evolution and Ethics; Science and Morals* (New York: Prometheus Books 2004), 122.
2. Edmund O. Wilson. *On Human Nature* (Boston: Harvard University Press,1978), 53.
3. Charles Krauthammer. Personal statement. *Human Cloning and Human Dignity: The Report of the President's Council on Bioethics* (New York: Public Affairs, Leon Kass, ed., 2002), 328.
4. Albert R. Jonsen. *The Birth of Bioethics* (New York: Oxford University Press, 1998), 348.
5. Robert P. George and Alfonso Gómez-Lobo. "The moral status of the human embryo." *Perspectives in Biology and Medicine*, 48 (2005):201-210.

6. Gilbert Meilander. "Some Protestant Reflections." In *The Human Embryonic Stem Cell Debate* (Cambridge: MIT Press, Suzanne Holland, Karen Lebacqz, and Laurie Zoloth, eds., 2001), 151.
7. Gilbert Meilaender. This statement and following statements from an interview with author. June 2005.
8. John Locke. *An Essay Concerning Human Understanding.* (Cleveland Ohio: World, A. D. Woosley, ed.,1964), 220.
9. Robert Song. "To be willing to kill a person." In *God and the Embryo: Religious Voices on Stem Cells and Cloning.* (Washington: Georgetown University Press, Brent Waters and Ronald Cole-Turner, eds., 2004), 102.
10. Leon L. Kass. *Life, Liberty and the Pursuit of Dignity: The Challenge for Bioethics.* (San Francisco: Encounter Books, 2002), 88.
11. Southern Baptist Convention. "Resolution on human embryonic and stem cell research." In *God and the Embryo*, 179-180.
12. Select Committee on Stem Cell Research. "A theologian's brief on the place of the human embryo within the Christian tradition, and the theological principles for evaluating its moral status." In *God and the Embryo*, 190-200.
13. John Haldane and Patrick Lee. "Aquinas on human ensoulment: abortion and the value of life." *Philosophy*, 78 (2003):255-278.
14. The Holy See declaration can be found at http://www.vatican.va/roman_curia/congregations/cfaith/documents/rc_con_cfaith_doc_19741118_declaration-abortion_en.html.
15. Prosaically described as the "Instruction on Respect for Human Life In its Origin and the Dignity of Procreation: Replies to Certain Questions of the Day" and found on the Vatican's website: http://www.vatican.va/roman_curia/congregations/cfaith/documents/rc_con_cfaith_doc_19870222_respect-for-human-life_en.html.
16. The Vatican's "Declaration on the Production and the Scientific and Therapeutic Use of Human Embryonic Stem Cells" can be

found at http://www.vatican.va/roman_curia/pontifical_academies/acdlife/documents/rc_pa_acdlife_doc_20000824_cellule-staminali_en.html.

17. Complete data of Harris Poll #58, August 18, 2004, is available at http://www.harrisinteractive.com/harris_poll/index.asp?PID=488.
18. Presbyterian Church (USA) Resolution. "Overture 01-50. On adopting a resolution enunciating ethical guidelines for fetal tissue and stem cell research—from the Presbytery of Baltimore." In *God and the Embryo*, 185-189.
19. Michael Mendiola. "Possible approaches from a Catholic perspective." *The Human Embryonic Stem Cell Debate* (Cambridge, MA: MIT Press, 2001), 124.
20. Ted Peters and Gaymon Bennett. "A plea for beneficence." *God and the Embryo*, 114-130.
21. Gaymon Bennett. Interview with author. August 2004.
22. Ted Peters and Gaymon Bennett. "Stem Cell Research and the claim of the other in the human subject." *Dialog: A Journal of Theology*, 43 (3) (2004):184-204.
23. Laurie Zoloth. This statement and following statements from an interview with author. June 2005.
24. Rabbi Elliot M. Dorff. *Ethical Issues in Human Stem Cell Research*, Vol. II (Rockville, MD: Religious Perspectives, 2000), C1. http://www.bioethics.gov/reports/past_commissions/index.html.
25. Rabbi Moshe Dovid Tendler. Ibid., C2.
26. Abulaziz Sachedina. Ibid., G1-3.
27. Mary Anne Warren can be consulted for her descriptive examples of sentience of animals in *Moral Status* (Oxford, Clarendon Press, 1997), 52-69.
28. Michael Gazzaniga. *The Ethical Brain* (Washington, DC: Dana Press, 2005), 7.
29. Ibid., 8.

30. James Q. Wilson. Personal statement. In *Human Cloning and Human Dignity*, 347-351.
31. See note 23.
32. See notes 21 and 22.
33. James C. Petersen. "Is the human embryo a human being?" *God and the Embryo*, 85.
34. Leon L. Kass. *Life, Liberty and the Pursuit of Dignity*, 4.
35. Ibid., 101.
36. Ibid., 103.
37. Ibid., 91.
38. The President's Council on Bioethics published a volume dedicated to the *Brave New World* philosophy in 2003. A copy can be obtained from http://bioethics.gov/reports/beyondtherapy/index.html.
39. Michael J. Sandel. "The case against perfection." *Atlantic Monthly*, April 2004. http://www.theatlantic.com/doc/ 200404/sandel.
40. Daniel Callahan. *What Price Better Health? Hazards of the Research Imperative* (University of California Press, October 2003), excerpts from Chapter 3 found at http://www.bioethics.gov/background/callahan_paper.html.
41. Francis Fukuyama, *Our Posthuman Society: Consequences of the Biotechnology Revolution* (New York: Farrar, Straus and Giroux, 2002), vii.
42. Francis Fukuyama. "Transhumanism." *Foreign Affairs: The Worlds Most Dangerous Ideas* (September/October 2004). http://www.foreignpolicy.com/story/cms.php?story_id=2696.
43. Francis Fukuyama. "Human biomedicine and the problem of governance." *Perspectives in Biology and Medicine*, 48 (2005):195-200.
44. Ibid.
45. Rebecca Dresser. This statement and the following statements from an interview with author. May 2005.
46. Alta Charo. This statement and the following statements from an interview with author. June 2005.

47. About 800 people attended the biggest convention of bioethics professionals in 2005. Over 2,000 people attended the 2005 meeting of the International Society of Stem Cell Research. The ISSCR's first meeting, held in 2002, attracted several hundred participants. Personal communication, Laurie Zoloth and ISSCR representatives.
48. Elizabeth Blackburn. "Thoughts of a former council member." *Perspectives in Biology and Medicine*, 48 (2005):173.
49. Ibid. 176.
50. Janet Rowley. Personal statement. In *Human Cloning and Human Dignity*, 340-342.
51. Arthur Caplan. "Is it ethical to use enhancement technologies to make us better than well?" *Public Library of Science Biology*, 1 (2004). http://medicine.plosjournals.org.
52. National Research Council. Guidelines for Human Embryonic Stem Cell Research (Washington, DC: National Academies Press, 2005), 43.
53. See note 52.
54. See note 43.
55. Henry Greely, et al. "Hi, I'm Mickey." Unpublished manuscript furnished by Henry Greely and the Stanford University Center for Biomedical Ethics.

CHAPTER 9

1. Robert Weinberg. "Of clones and clowns." *Atlantic Monthly* (June 2002). http://www.theatlantic.com/doc/200206/ weinberg.
2. In 1973, the Department of Health, Education and Welfare (DHEW) established a moratorium on federally funded research on live fetuses. In 1974, Congress adopted a similar moratorium, including in the ban human embryos created by IVF. In 1975, DHEW regulations superseded the moratoria. Incorporated into the Code of Federal Regulations (CFR), they prohibited federal

funding of IVF experimentation unless approved by an Ethics Advisory Board. Such a board did find some experiments in this arena "ethically acceptable" in a 1979 report, but the secretary of DHEW did not approve funding, the Board expired, and no human embryo experiments were performed with federal funds.

3. See Leon Kass. "Ethical issues in human in vitro fertilization, embryo culture and research, and embryo transfer," In Vitro Fertilization, Appendix (Ethics Advisory Board, U.S. Department of Health, Education and Welfare, May 4, 1979); and "Making babies' revisited," *Public Interest*, 54 (Winter, 1979):32-60.
4. Albert Jonsen. *The Birth of Bioethics* (New York: Oxford University Press, 1998), 311.
5. Summarized in Dorothy C. Wertz. "Embryo and stem cell research in the United States: history and politics," *Gene Therapy*, 9 (2002): 674-678.
6. National Institutes of Health Embryo Research Panel Report, Vol. I, September 1994. http://www.bioethics.gov/reports/past_commissions/index.html.
7. The National Bioethics Advisory Commission. *Ethical Issues in Stem Cell Research*, Vol.I. http://www.bioethics.gov/reports/past_commissions/index.html.
8. John C. Fletcher presents one summary of the NBAC's ethical arguments in "The stem cell debate in historical context." *The Human Embryonic Stem Cell Debate* (Cambridge, MA: MIT Press, Suzanne Holland, Karen Lebacqz, and Laurie Zoloth, eds., 2001), 61-72.
9. Excerpts in this chapter come from the author's published essays, including "The consequences of the restrictions of human stem cell research" (with co-author Tom Maeder), *Acumen Journal of Sciences* 1(1):36-45, and "Ethical dilemma." *Acumen Journal of Sciences*, 1 (2): 66-70.
10. Eugene Russo. "Advice fit for a president." *The Scientist*, 16 (4):22.

11. Paul Berg. Interview with the author. October 2004.
12. White House press release. "President Bush calls on Senate to back human cloning ban." http://www.whitehouse.gov/news/releases/2002/04/20020410-4.html.
13. Orrin Hatch. Letter to the president, March 2, 2004. http://democrats.reform.house.gov/features/politics_and_science/example_stem_cells.htm.
14. Sheryl Gay Stolberg. "Limits on stem cell research re-emerge as a political issue." *The New York Times* (May 6, 2004), A1.
15. Department of Heath and Human Services press release. July 14, 2004. http://www.hhs.gov/news/press/2004pres/20040714b.html.
16. John Kerry's position statement. "Supporting stem cell research to find cures for millions of Americans suffering from debilitating diseases." Originally from his campaign website.
17. John C. Danforth. "In the name of politics." *The New York Times* (March 30, 2005), A17.
18. Sheryl Gay Stolberg. "Sponsor of stem cell bill says senate could override a veto." *The New York Times* (May 26, 2005), A25.
19. Cheryl Gay Stolberg. "Senate leader veers from Bush over stem cells." *The New York Times* (July 28, 2005), A1.
20. Albert Jonsen. Interview with the author. March 2003.
21. The moral positions of individual council members are fluid, underscoring how time and debate can modify positions. Francis Fukuyama voted for the moratorium in 2002 but on the eve of the California Stem Cell Initiative vote in October 2004 wrote in *The Wall Street Journal*, "I believe that President Bush's current stem cell policy is much too restrictive." Michael Sandel voted against the moratorium and for therapeutic uses of stem cells, but in an Atlantic Monthly article published in April 2004, compared the use of American biotechnological advances to modern eugenics. Both commentators share the view that science is underregulated. See

Michael Sandel. "The case against perfection," *Atlantic Monthly* (April 2004), and Francis Fukuyama. "Big Science, Big Giveaway." *The Wall Street Journal* (October 25, 2004), A18.

22. Donald Kennedy. "An epidemic of politics." *Science*, 299 (2003):62.
23. Julie Wakefield. "Science's political bulldog." *Scientific American*, (May 2004): 50-52.
24. For more on the controversy surrounding scientific appointments, see "Scientific integrity in policymaking: an investigation into the administration's misuse of science." Union of Concerned Scientists (March 2004). http://www.ucsusa.org/global_environment/rsi/page.cfm?pageD=1642.
25. Editorial. "Does Washington Really Know Best?," *The Lancet*, 364 (2004):114.
26. Andrew C. Revkin. "Bush vs. the Laureates: How Science Became a Partisan Issue." *The New York Times* (Oct 19, 2004), F1.
27. Alden Myer is quoted in a report from the Union of Concerned Scientists. "Restoring Scientific Integrity." http://www.ucsusa.org/global_environment/rsi/page.cfm?pageID=1472.
28. Janet D. Rowley, et al. "Harmful moratorium on stem cell research." *Science*, 297 (2002):1957.
29. Elizabeth Blackburn and Janet Rowley. "Reason as our guide." *Public Library of Science Biology*, 2(4) (2004): e116 DOI: 10.1371/journal.pbio.0020116.
30. From the Union of Concerned Scientists. "New cases of scientific abuse by administration emerge: thousands more scientists join protest." http://www.ucsusa.org/news/press_release.cfm?newsID=405.
31. Rick Weiss. "Bush ejects two from bioethics council." *Washington Post* (February 28, 2004), A6.
32. Ibid.
33. Claudia Kalb and Debra Rosenberg. "The battle over stem cells." *Newsweek* (October 25, 2004):46-47.

34. James Battey. "Update on the status of the 78 eligible entries on the NIH human embryonic stem cell registry as of February 23, 2004." http://olpa.od.nih.gov/hearings/108/session1/reports /hseap-prop.asp.
35. Connie Bruck. "The hollywood science." *New Yorker* (October 18, 2004):67.
36. See note 11.
37. Rick Weiss. "Approved stem cells potential questioned." *Washington Post* (October 29, 2004), A3.
38. Ibid.
39. Rick Weiss. "NIH agency chiefs criticize federal policy on stem cells." *Washington Post* (April 7, 2005), A29.
40. The author attended several meetings during the writing of this book, interviewed presenters, and surveyed the abstracts and poster sessions attached to each event.
41. As of November 2, 2004, just 22 lines on the NIH registry came from American laboratories.
42. The International Society of Stem Cell Research tracks the availability of unpublished and published hESC lines. http://www.isscr.org/science/sclines.htm.
43. Ted Peters and Gaymon Bennett. "A plea for beneficence." *God and the Embryo: Religious Voices on Stem Cells and Cloning* (Washington, DC: Georgetown University Press, Brent Waters and Ronald Cole-Turner, eds., 2003), 118.
44. Quirin Schiermeier. "Divergent local law threaten to stifle Europe's stem cell project." *Nature*, 434 (2005):809.
45. Associated Press. "Brazil funds human stem cell research." http://news.ninemsn.com.au/article.aspx?id=48698.
46. Steven Minger. "Investment in life." *The Guardian Unlimited* (May 20, 2004). http://www.guardian.co.uk/life/interview/story/0,12982,1220142,00.html.

47. Aaron Levine. "Trends in the geographic distribution of human embryonic stem-cell research." *Politics and Life Sciences*, 23 (2005): 40-45.
48. Warren Hoge. "U.S. stem cell policy delays U.N. action on human cloning." *The New York Times* (October 24, 2004), Section 1, 8.
49. CBS News Sunday Morning. "Stem cell research's wide divide." Broadcast June 5, 2005.
50. Because Melton's cultures were created after August 1, 2001, they are not eligible for NIH funding and will not be posted on the government's registry.
51. Rick Weiss. "Harvard team seeks to clone embryos for stem cells." *Washington Post* (October 14, 2004), A1.
52. NIH budget request for 2006. http://www.nih.gov/about/director/budgetrequest/fy2006directorsbudgetrequest.htm.
53. Robert Klein. Interview with author. February 2005.
54. Irving Weissman. Interview with author. June 2003.
55. Paul Berg. Interview with author. October 2004.
56. Private interview, June 2004.
57. See note 1.
58. Christopher Scott and Thomas Maeder. "The consequences of the restrictions of human stem cell research." *Acumen Journal of Sciences*, 1(1) (2003):36-45.
59. Survey results found at http://www.harrisinteractive.com.
60. Survey conducted by Opinion Research Corporation (ORC) on behalf of the Results for America (RFA) project of the nonprofit and nonpartisan Civil Society Institute. http://www.resultsforamerica.org
61. Survey results found at http://www.camradvocacy.org.

CHAPTER 10

1. The DNA of the cloned cells is identical, but the expression of the genes is different from the original somatic cell. See Chapter 4.
2. Jose B. Cibelli, et al. "Somatic cell transfer in humans: pronuclear and early embryonic development." *Journal of Regenerative Medicine,* 2 (2001). http://65.205.1.226 cloning/human-clones1.html.
3. Carina Dennis. "Chinese fusion method promises fresh route to human stem cells." *Nature,* 424 (2003):711.
4. Woo Suk Hwang, et al. "Evidence of a pluripotent human embryonic stem cell line derived from a cloned blastocyst." *Science,* 303 (2004):1669–1674.
5. Woo Suk Hwang, et al. "Patient-specific embryonic stem cells derived from human SCNT blastocysts." *Science,* 238 (2005):1777–1780.
6. David Magnus and Mildred Cho. "Issues in oocyte donation for stem cell research." *Science,* 308 (2005):1747–1748.
7. Seoul National University does not provide an English version of the report on their website, but a summary text can be found on the Aljazeera news website: http://english.aljazeera.net/NR/exeres/C1096296-2CCE-4E9D-A170-C5EA56DE2F5E.htm
8. Editorial, "Study shows bioethics awareness lacking." Dong-a Ilbo (East Asia Daily, Seoul) 25 Nov 2005; http://english.donga.com/srv/service.php3?biid=2005112564428.

EPILOGUE

1. Francois Jacob. *The Possible and the Actual* (New York, Pantheon Books, 1982): 61.
2. Patricia Anstett. "Stem cell research activist, Dr. Wise Young, tackles current issues." *Detroit Free Press* (June 17, 2005), 1.
3. Ibid.

4. Wise Young's website tracks clinical trials for spinal injury. http://carecure.atinfopop.com.
5. Bruce Carlson. *Human Embryology and Developmental Biology*, Third Edition (St. Louis: C.V. Mosby, 2004), 270-274.
6. Carl Zimmer. "Michael Gazzaniga; a career spent learning how the mind emerges from the brain." *The New York Times* (May 10, 2005), F3.
7. Ernlé Young. "Stem cell research: therapeutic possibilities and ethical controversies." Unpublished lecture transcript (2003), Monash University, Melbourne, Australia.

Glossary

adult stem cell Stem cell found in different tissues and organs of the developed animal. Adult stem cells can both renew themselves (make more stem cells by cell division) and differentiate (divide and, with each cell division, evolve more and more into different types of cells).

allogeneic transplant Cell, tissue, or organ transplants from one member of a species to a genetically different member of the same species.

amino acid Any one of 20 molecules that serve as building blocks for proteins.

antigens Protein marker on a cell surface. Antigens can be foreign substances (chemicals, bacteria, viruses, or pollen) that cause the immune system to produce antibodies against it. *See surface antigens.*

astrocyte Star-shaped brain cell that provides structural support and is implicated in learning and memory.

asymmetric cell division Type of cell division peculiar to stem cells. The two daughter cells produced are different. One becomes a stem cell exactly like the parent, the other becomes a more mature cell.

autoimmune disease Condition in which the body recognizes its own tissues as foreign and mounts an immune response against them.

autologous transplant After a medical treatment, a cell or bone marrow transplant from one individual back to the same individual. Such transplants do not induce an immune response and are not rejected.

base Chemical "letters" of DNA or RNA. The four bases in DNA are adenine (A), thymine (T), guanine (G), and cytosine (C). In RNA, uracil (U) replaces thymine. *See nitrogenous base.*

blastocyst Early human embryo consisting of approximately 100 cells. It contains a fluid-filled cavity; an internal cluster of cells, the inner cell mass; and an outer layer of cells, the trophoblast. Also called the early or pre-implantation embryo.

blood chimera Person with two genetically distinct types of blood cells. Natural blood chimeras can be non-identical twins who shared their blood supply in the uterus.

bone marrow transplant Procedure to replace bone marrow destroyed by treatment cancer or radiation. Transplantation can be autologous (an individual's own marrow saved before treatment) or allogeneic (marrow donated by someone else).

central nervous system (CNS) Information-processing organ of the nervous system, consisting of the brain, spinal cord, and motor neurons.

chromosome DNA molecule that carries genes, the hereditary material of an organism.

cleavage Specialized kind of cell division where a single-celled zygote subdivides into many smaller cells without increasing in size.

clinical trials Research study in human volunteers designed to answer specific health or medical questions.

clone Genetically identical population of organisms, cells, viruses, or DNA. *See cloning.*

cloning Cloning may occur by propagation of cuttings, as in the case of plants; budding, as in the case of hydra; fission, as in the case of bacteria and protozoa; or by somatic cell nuclear transfer, as in the case of higher-order animals, such as mammals. Cloning can also be applied to a group of cells undergoing division.

cord stem cell Hematopoietic stem cell present in the blood of the umbilical cord during and shortly after delivery.

co-transplantation Hypothetical therapy replacing the bone marrow of a patient with compatible immune cells made from an embryonic stem cell line. Other cells, tissue, and organs can be transplanted into the patient with no rejection because both the immune cells and other tissues come from the same genetic source.

crossing over Interaction between paired chromosomes during meiosis, whereby portions of chromosomes—and genetic information—is exchanged.

cytoplasm Jelly-like insides of the cell bound by the plasma membrane.

daughter cell Any cell that results from the division of a single cell.

dedifferentiation Regression to a younger cell type along a pathway of differentiation. SCNT is an example of dedifferentiation.

developmental biology Discipline that studies embryonic and other developmental processes.

differentiation Process of development accompanied by an increase in the level of organization, complexity, or specialization of a cell or tissue.

DNA (deoxyribonucleic acid) Molecule made of four chemical units called bases (abbreviated A, G, C, and T) that contain genetic information (genes) encoded in the sequence of bases.

dopamine Neurotransmitter formed in the brain essential to the normal functioning of the central nervous system. A reduction in its concentration is associated with Parkinson's disease.

ectodermal layer The outer of three germ layers of the early embryo that develops into skin, cells of the amnion and chorion, nervous system, enamel of the teeth, lens of the eye, and neural crest.

embryo The product of a fertilized egg, from the zygote until the fetal stage.

embryoid bodies Spherical colonies of embryonic stem cells seen only in culture, containing cells of all three germ layers: endoderm, mesoderm, and ectoderm.

embryogenesis Development and formation of an embryo.

embryology Branch of biology that studies the early development and formation of living organisms

embryonal carcinoma cell (EC cell) Stem cell cultured from the inside of a teratoma can produce a line of EC cells. EC cells, although cancerous, are similar to embryonic stem cells.

embryonic stem cell (ESC) Cell cultured from the inner cell mass of developing blastocysts. An ESC cell is self-renewing (can replicate), pluripotent (can form all cell types found in the body), and can live indefinitely.

endodermal layer The inner of three germ layers of the early embryo that develops into lungs, the intestine, the liver, and the pancreas.

enucleated cell Cell whose nucleus has been removed.

enzyme Protein that speeds up a specific chemical reaction.

epidermis Outer layers of the skin.

epigenetic Chemical change to DNA that results in a change of a gene's expression. Although a mutation permanently changes the order of DNA (any of the chemical letters A, G, C or T), an epigenetic change does not. Epigenetic changes may be reversible.

feeder layer Layer of cultured living, non-dividing cells that support the growth of a stem cell line.

fertilization Fusion of the egg and sperm, each of which contains half the required number of chromosomes. The union produces a single-celled zygote, the earliest form of embryo.

fetus In humans, the developmental stage starting from the end of the embryonic stage, 7–8 weeks after fertilization, to birth.

fluorescence activated cell sorter (FACS) Scientific instrument used to measure the characteristics of individual cells by sorting them based on the amount of light emitted by each cell. FACS is also used to purify stem cells.

gametes Mature egg (oocyte) or sperm (spermatocyte); cells that carry genes to the next generation.

gene Segment of DNA consisting of a unit of inherited information.

gene expression Process of translating the information in a gene in order to produce a protein. Expression requires transcription of a gene's DNA sequence into RNA and translation of the RNA sequence into a protein.

gene therapy Experimental technique using genes to treat or prevent disease. In the future, this technique may allow doctors to treat a disorder by inserting a gene into a patient's cells instead of using drugs or surgery.

genetic code The "dictionary" used to decipher which sequence of chemical bases in DNA is used to produce a specific trait. Triplets of chemical bases code for 20 amino acids, which are the building blocks of proteins.

genetic diversity Variation and complexity of individual genes.

genome All the genetic information in a cell, including genes and other DNA sequences.

germ layers The three germ layers (endoderm, mesoderm, and ectoderm) of the early embryo that give rise to all tissues of the body.

graft-versus-host-disease (GVHD) Sometimes-fatal reaction of donated immune cells against the patient's tissue.

granulocyte colony stimulating factor (G-CSF) Protein that stimulates the production of stem cells.

hematopoietic stem cell (HSC) Stem cells that make red blood cells, white blood cells, and platelets. They live within the bone marrow but are also found in umbilical cord blood, peripheral blood, and in the liver.

hematopoietic stem cell transplantation Transplantation of stem cells with blood-forming potential.

HLA (histocompatibility antigens) Antigens found on the surface of nearly every cell in the body, including white blood cells. The body's immune system uses HLA antigens to recognize "self" from "nonself" (foreign substances). Tissues from most people have little HLA matching to others. In siblings, the probability of compatibility is higher, whereas identical twins are almost always compatible.

human embryonic stem cell (hESC) line Embryonic stem cells maintained and grown in culture.

immune cells White blood cells that form in the bone marrow. They include T and B lymphocytes, among many others.

in vitro Latin for "under glass," the reference to biological research conducted in a laboratory. *See in vivo.*

***in vitro* fertilization (IVF)** Laboratory procedure where an egg cell (oocyte) is fertilized by sperm cells in a plastic dish (i.e., *in vitro*). The resulting fertilized egg, called a zygote, will start dividing and, after a several divisions, forms the embryo that is implanted into the womb.

in vivo Latin for "in a living body," the reference to biological research conducted in an animal. *See in vitro.*

inner cell mass (ICM) Cluster of cells along the inner wall of the blastocyst. Culturing the cells that make up the ICM makes embryonic stem cells.

lipoprotein Fatty packages of cholesterol that travel through the blood stream.

meiosis Special type of cell division by which produces the eggs and sperm. Four cells are produced, each containing half the number of chromosomes.

mesenchymal stem cell (**MSC**) Mixed population of cells coming from the non blood-forming regions of bone marrow and capable of self-renewal and differentiation into cells from all three germ layers.

mesodermal layer The middle of three germ layers of the early embryo that develops into muscle, bone, and blood.

mitosis Central feature of cell division that includes duplication of identical copies of chromosomes.

morphology Study of the shape and visual appearance of cells, tissues, and organs.

morula Early mammalian embryo resembling a mulberry, usually a solid ball of cells.

multipotent Stem cells that make multiple differentiated cell types, but within a particular tissue, organ, or physiological system. For example, multipotent blood or hematopoietic stem cells (HSC) can produce all cell types that are components of the blood.

multipotent adult progenitor cells (**MAPCs**) Adult stem cells with apparent embryonic stem cell properties.

neural stem cell (**NSC**) Stem cell of the brain, which can make new neurons, astrocytes, and oligodendrocytes.

neurons Elongated nerve cell that receives, conducts, and transmits signals in the nervous system.

niche Microenvironment in the body where stem cells normally reside.

nitrogenous base (**base pair or base**) The chemical "letters" of DNA or RNA. The four bases in DNA are adenine (A), thymine (T), guanine (G), and cytosine (C). In RNA, uracil (U) replaces thymine.

nuclear transfer *See somatic cell nuclear transfer.*

nucleus Membrane bound sac in the cell that contains DNA, which is organized into chromosomes.

oligodendrocyte Cell of the central nervous system that forms myelin, a fatty material essential for the conduction of nerve signals.

oocyte Primitive cell of the human fetus that develop into eggs at puberty.

organogenesis Development of specific organs, such as the lungs, heart, and eyes.

parthenogenesis Form of reproduction in which an egg develops without the fusion of sperm with the egg cell. Artificially inducing parthenogenesis with human eggs may be a means to isolate stem cells from an embryo.

placenta Spongy tissue in the uterus from which the embryo gets nourishment and oxygen.

plasma membrane Thin membrane surrounding a living cell. The membrane allows certain small molecules to pass back an forth but acts as a barrier to others.

plasticity Phenomenon used to describe a stem cell that seems to be capable of becoming a specialized cell type not originally associated with its germ layer or switching its cell lineage. *See transdifferentiation* and *dedifferentiation*.

pluripotent Stem cells that make all of the cell types in found in an embryo, fetus, or developed organism (but not the trophoblast and placenta). Embryonic stem cells are considered pluripotent.

Preimplantation Genetic Diagnosis (PGD) Genetic analysis of a human embryo made in an IVF clinic. A single cell is removed from the embryo and tested. If the embryo has no genetic disorders, it is placed into the uterus.

preimplantation embryos Fertilized egg (zygote) and all the developmental stages up to, but not beyond, the blastocyst stage.

President's Council on Bioethics (PCBE) Group of ethicists and scholars that address the ethical and political effects of biomedical innovation.

progenitor cell Early descendant of a stem cell that can only differentiate but not self-renew. A progenitor cell is often more limited in the kinds of cells it can become than a stem cell.

proteins Molecules composed of chains of amino acids in a specific order, determined by the sequence of bases (A, T, G, and C) in the gene. Proteins are everywhere in the body and are necessary for the structure, function, and regulation of cells, tissues, and organs. Examples are hormones, enzymes, and antibodies.

regenerative medicine Medical interventions using stem cells that aim to repair and replace damaged or diseased organs.

reproduction Creation of a new individual from two parents, usually by the union of an egg and sperm.

reproductive biology Study of reproduction.

reproductive cloning Term for making live animals using somatic cell nuclear transfer (SCNT). A nucleus from the cell of a donor animal is put into an empty egg and grown in culture into a blastocyst. The embryo is implanted into the uterus of a surrogate mother. If the embryo survives to term, the newborn is technically "a genetic clone" of the animal that donated the nucleus. *See somatic cell nuclear transfer.*

reprogramming Erasing certain chemical changes to DNA without changing the genes themselves (the order of chemical letters A, T, G and C). Somatic cell nuclear transfer is the best example of reprogramming at work. Genes of a somatic cell nucleus are reprogrammed (but not rearranged) in order to make an embryo. *See epigenetic.*

ribosome Structure in the cytoplasm essential for the manufacture of proteins.

RNA (ribonucleic acid) Molecule made of four chemical letters called bases (abbreviated A, G, C and U). U substitutes for the T of DNA.

self-renewal Stem cell division that produces a daughter cell just like the mother cell. The other daughter cell is born different and becomes a cell more restricted in its potential. *See aymmetric cell division.*

Severe Combined Immunodeficiency (SCID) Also called the bubble boy disease, SCID is a group of very rare, life-threatening diseases that are present at birth. SCID children have little or no immune systems.

sexual reproduction Fertilization of an egg by a sperm, sometimes followed by smoking a cigarette and staring at the ceiling.

somatic cell nuclear transfer (SCNT) Laboratory technique in which the nucleus of a somatic cell (any cell of the body except sperm and egg cells) is injected, or transferred, into an egg that has had its nucleus removed.

somatic cells Any cell within the developing or developed organism with the exception of egg and sperm cells.

spermatocyte A cell that makes sperm cells.

stem cells Cell that has the capacity to both self-renew (make more stem cells by cell division) and differentiate into mature, specialized cells.

surface antigens Cell-specific protein that serves as receptor for incoming signals and for transporting amino acids across the cell membrane.

telomerase Enzyme active in cancer and actively dividing cells.

teratoma Non-lethal, naturally occurring tumor of the testes and ovaries.

terminally differentiated Non-dividing cell at the end of a pathway of cell differentiation. For example, a skin cell is terminally differentiated.

therapeutic cloning Laboratory technique that uses somatic cell nuclear transfer (SCNT) to make an hESC line. In theory, therapeutic cloning can produce any cell, organ, or tissue in the body. The cells are genetically identical to the patient and, if transplanted, will not be rejected by the patient's immune system.

tissue engineering Techniques used to grow skin, cartilage, bone, vessels, and parts of organs by manipulating cells in the laboratory.

totipotent Cells that can make any organ or tissue. The zygote and cells made immediately following fertilization are considered totipotent and can make any cell of the embryo or placenta.

trait Observable physical characteristics of a living thing. Traits are determined by proteins.

transdifferentiation Apparent ability of a particular cell of one tissue, organ, or system, including stem or progenitor cells, to switch into a cell type characteristic of another tissue, organ, or system—for example, blood stem cells that change to liver cells. *See plasticity*.

transgenic mouse Mouse whose genome has been altered by the introduction of new sequences of DNA. The genetic changes are heritable and are thus passed on to subsequent generations.

translation Process by which the genetic information contained in a chain of RNA is "decoded" into a protein with a specific sequence of amino acids.

trophoblast Cells surrounding a developing embryo that are responsible for implantation and formation of the placenta.

zygote Fertilized egg produced by the fusion of an egg and a sperm and the earliest form of the embryo.

Further Reading

The texts mentioned here represent a cross section of the themes presented in this book and are excellent resources for serious study of the many dimensions of stem cells. A few texts are only available through government websites. In those cases, Internet addresses are supplied after the titles.

GENETICS AND CELL BIOLOGY

Alberts, Bruce, et al. *Molecular Biology of the Cell,* Fourth Edition. New York: Garland Science. 2002.

Hartwell, Leland, et al. *Genetics: From Genes to Genomes,* Second Edition. New York: McGraw-Hill. 2003.

DEVELOPMENTAL BIOLOGY AND HUMAN EMBRYOLOGY

Carlson, Bruce M. *Human Embryology and Developmental Biology,* Third Edition. Philadelphia: Mosby. 2004.

Gilbert, Scott. *Developmental Biology,* Seventh Edition. Sunderland, MA: Sinauer. 2003.

Moore, Keith L., and T. V. M. Persaud. *The Developing Human: Clinically Oriented Embryology,* Seventh Edition. Philadelphia: Saunders. 2003.

Wolpert, Lewis, et al. *Principles of Development,* Second Edition. Oxford: Oxford University Press. 2002.

STEM CELL BIOLOGY

Carrier, Ewa, and Gracy Ledingham. *100 Questions and Answers About Bone Marrow and Stem Cell Transplantation*. Sudbury, MA: Jones & Bartlett. 2004.

Chiu, Arlene Y., and Mahendra S. Rao. *Human Embryonic Stem Cells*. Totowa, NJ: Humana Press. 2003.

Kiessling, Ann, and Scott Anderson. *Human Embryonic Stem Cells*. Sudbury, MA: Jones and & Bartlett. 2003.

Marshak, Daniel R., et al. *Stem Cell Biology*. Cold Spring Harbor, NY: Cold Spring Harbor Press. 2001.

President's Council on Bioethics. *Monitoring Stem Cell Research*. Washington, DC: President's Council on Bioethics. 2004a. http://bioethics.gov/reports/stemcell/index.html.

POPULAR SCIENCE

Berg, Paul, and Maxine Singer. *Dealing with Genes: The Language of Heredity.* Mill Valley, CA: University Science Books. 1992.

Colata, Gina. *Clone: The Road to Dolly, the Path Ahead*. New York: William Morrow. 1998.

Gazzaniga, Michael S. *The Ethical Brain*. New York: Dana Press. 2005.

Gordon, Jon W. *The Science and Ethics of Engineering the Human Germ Line: Mendel's Maze.* Hobokoen, NJ: Wiley-Liss. 2003.

Maienschein, Janet. *Whose View of Life? Embryos, Cloning and Stem Cells.* Cambridge, MA: Harvard University Press. 2003.

Parson, Ann B. *The Proteus Effect: Stem Cells and Their Promise for Medicine.* Washington, DC: Joseph Henry Press. 2004.

West, Michael D. *The Immortal Cell: One Scientist's Quest to Solve the Mystery of Aging.* New York: Doubleday. 2003.

ETHICS AND POLICY

Beauchamp, Tom L., and James F. Childress. *Principles of Biomedical Ethics*, Fourth Edition. New York: Oxford University Press. 1994.

Dworkin, Ronald. *Life's Dominion: An Argument About Abortion, Euthanasia and Individual Freedom*. New York: Vintage Books. 1993.

Holland, Suzanne, Karen Lebacqz, and Laurie Zoloth, eds. *The Human Embryonic Stem Cell Debate: Science, Ethics and Public Policy*. Cambridge MA: The MIT Press. 2001.

Huxley, Thomas. *Evolution and Ethics; Science and Morals*. New York: Prometheus Books. 2004.

Jonsen, Albert R. *The Birth of Bioethics*. New York: Oxford University Press. 1998.

National Research Council. *Guidelines for Human Embryonic Stem Cell Research*. Washington, DC: National Academies Press. 2005. http://www.nap.edu.

———. *Scientific and Medical Aspects of Human Reproductive Cloning*. Washington, DC: National Academies Press. 2002. http://www.nap.edu.

President's Council on Bioethics. *Beyond Therapy: Biotechnology and the Pursuit of Happiness*. Washington, DC: President's Council on Bioethics. 2003a. http://bioethics.gov/reports/beyondtherapy/index.html.

———. *Human Cloning and Human Dignity: an Ethical Inquiry*. Washington, DC: President's Council on Bioethics. 2003b. http://bioethics.gov/reports/cloningreport/index.html.

Steinbock, Bonnie. *Life Before Birth*. New York: Oxford University Press. 1992.

Warren, Mary Anne. *Moral Status: Obligations to Persons and Other Living Things*. Oxford: Clarendon Press. 1997.

Waters, Brent, and Ronald Cole-Turner, eds. *God and the Embryo*. Washington, DC: Georgetown University Press. 2003.

Acknowledgments

Thank you.

To the good people at Pi Press: Stephen Morrow and Jeffrey Galas. I am also grateful for friends and professionals who helped along the way, including Ben Roberts, Hiliar Chism, Luana Richards, Emi Quade, Paul Quade, and Mark Ong.

To the expert readers and reviewers of the manuscript:

Monya Baker, Paul Berg, Ben Bowen, Alta Charo, Alan Davidson, Coleen Dooley, Scott Dylla, Margaret Eaton, Donald Kennedy, Dan Erlansen, Mike Gazzaniga, Aaron Hirsch, David Hirsch, Jennifer Knight, David Magnus, Douglas Melton, Susie Prohaska, Ben Roberts, Anne Scott, Ernlé Young, Irving Weissman, Laurie Zoloth, and Leonard Zon.

To the enthusiastic and passionate professionals interviewed for the book:

Michael Abeyta, Anthony Atala, Mark Beer, Paul Berg, Gaymon Bennett, Anders Bjorklund, Alta Charo, Mildred Cho, Rebecca Dresser, Scott Dylla, Ted Golos, Michael Gazzaniga, Zach Hall, Joshua Hare, Rudolph Jaenisch, Albert Jonsen, Regis Kelly, Donald Kennedy, Robert Klein, Diana Krause, Robert Lanza, Bertram Lubin, Gilbert Meilaender, Tom Okarma, Susie Prohaska, Matthew Scott, Bernard Siegel, Saul Sharkis, Ralph Snodgrass, James Thomson, Alan Trounson, Ann Tsukamoto, Yury Valensky, Amy Wagers, David Wininger, Irving Weissman, Ernlé Young, and Laurie Zoloth.

Especially:

Paul Berg provided the spark for the project. Hundreds of graduate students and postdoctoral fellows are grateful to Paul for his scientific passion, uncompromising attention to detail, and flair for clear explanation. I am, too.

Susie Prohaska was a godsend, taking time out from experiments, lab meetings, and grant writing to help research the book. My files are swollen with papers she sent—solicited or otherwise.

Ernlé Young guided the manuscript through its evolution and read every word, sometimes twice. I appreciate his calm and evenhanded approach to the difficult chapters.

Stephen Meyer of the Stem Cell Action Network (SCAN) provided me—and hundreds of other subscribers—with daily news feeds about stem cells.

There would be no book without Linda Paulson. She said, simply, "Write it." Under her tutelage, the Stanford University MLA Works-in-Progress group provided input into numerous chapters. They are Aline Beck, Bill Beck, Jennifer Burton, George Coomb, Walt Cook, Gail Graham, and Beth Karpas. The imprints of Mark North and Stuart Wells are found throughout the back chapters.

Finally, and most importantly, a note to my family—a tribe of writers, artists, inquisitors, and stern grammarians—it all comes to roost somehow, some early, some late. Questions about "The Book" came first, even when I wished for a different topic. Their energy and interest fueled days when I was certain not a word would appear.

Index

Page numbers followed by *n* signify endnotes.

W

X-Y-Z